Acting in Time
on Energy Policy

Acting in Time on Energy Policy

Kelly Sims Gallagher
editor

BROOKINGS INSTITUTION PRESS
Washington, D.C.

ABOUT BROOKINGS

The Brookings Institution is a private nonprofit organization devoted to research, education, and publication on important issues of domestic and foreign policy. Its principal purpose is to bring the highest quality independent research and analysis to bear on current and emerging policy problems. Interpretations or conclusions in Brookings publications should be understood to be solely those of the authors.

Copyright © 2009
THE BROOKINGS INSTITUTION
1775 Massachusetts Avenue, N.W., Washington, D.C. 20036
www.brookings.edu

Library of Congress Cataloging-in-Publication data
Acting in time on energy policy / Kelly Sims Gallagher, editor.
    p.   cm.
    Includes bibliographical references and index.
    Summary: "Details the often incomplete, incoherent, and ineffectual U.S. energy policy, tackling specific policy questions: Why are these components of energy policy so important? How would 'acting in time'—not waiting until politics demands action—make a difference? What should our policy actually be? Proposes measures to overcome this counterproductive tendency"—Provided by publisher.
    ISBN 978-0-8157-0293-1 (pbk. : alk. paper)
    1. Energy policy—United States.   I. Gallagher, Kelly Sims.   II. Title.
    HD9502.U62A37 2009
    333.790973—dc22 2009009054

9 8 7 6 5 4 3 2 1

The paper used in this publication meets minimum requirements of the American National Standard for Information Sciences—Permanence of Paper for Printed Library Materials: ANSI Z39.48-1992.

Typeset in Minion and Univers Condensed

Composition by R. Lynn Rivenbark
Macon, Georgia

Printed by R. R. Donnelley
Harrisonburg, Virginia

Contents

Foreword

David T. Ellwood

The question of whether we can "act in time" on energy and climate change poses one of the most profound challenges facing the world today. No human activity, other than the wide-scale use of nuclear weapons, has greater potential to reshape and harm our planet and our species than the rapidly expanding generation of greenhouse gases. What is so frustrating about the issue is that even though the dangers are widely accepted in the scientific community, and even though failing to act in time could set off a chain of events that would be all but irreversible, action to date has been weak at best.

"Acting in time" has been a theme of a larger research project housed at the Harvard Kennedy School. It focuses precisely on the question of why nations, institutions, and individuals so often seem unwilling or unable to act in time, even when problems are easily seen coming and even when acting sooner rather than later will be far less costly in the long run.

The project was stimulated in part by Hurricane Katrina and the devastation of New Orleans. This event is typically described as an enormous failure in response. And so it was. It took days and even weeks to even begin providing real aid. But the greater tragedy of Hurricane Katrina was that everyone knew it would happen. In New Orleans, one can walk down streets and see ocean liners going by *above* street level. Even the most casual observer understands that he or she is standing in a bathtub and that the walls of this tub are not very high. There were many, many studies that suggested that New

Orleans would be devastated by even a moderate hurricane, that the levees needed to be significantly improved, and that larger environmental changes were vital. National agencies listed a hurricane in New Orleans as one of the three greatest risks facing the United States, yet no one acted in time. Now we are spending vastly more billions to rebuild than we would have needed to spend in the first place to prevent the tragedy. And some wonder whether the city will ever recover its former vitality.

There are plenty of other examples. Consider what is happening with Social Security or Medicare. The combination of demographic changes and rising medical costs absolutely guarantees that these two vital programs will run out of money, forcing drastic increases in taxes or cuts in benefits unless we reform them quickly. The only actions taken so far—to expand Medicare benefits—have made the problem worse. Along with pandemics (such as avian flu), nuclear terrorism, and natural disasters, these are "predictable surprises," to use Max Bazerman's term.

At this writing, the world is preoccupied with a financial crisis. Perhaps this, too, was a predictable surprise—though far fewer people saw this collapse coming. Now, when it is too late, we scramble to act. And in our preoccupation with this immediate crisis, we may once again simply ignore the crises sure to come. Many claim that climate change is on the front burner, but will our leaders really be able to face the difficult choices serious action will entail, especially if the world is in recession? This book explores what might be done, what might prevent action from being taken, and what might spur further action.

There are a number of lessons derived from our larger project at the Kennedy School that bear directly on the energy and climate change issue. We define "acting in time" problems as cases where people fail to act to prevent or lessen some future crisis, cost, or catastrophe that is easily predictable and relatively well understood, and where acting sooner rather than later would have a very high payoff. We have also been able to identify situations where inaction is particularly likely. And honestly, the energy and environment crisis has more action-stopping characteristics than virtually any other problem we have seen. I will highlight just a few of these factors, but sadly there are plenty more. However, I will try not to leave you in despair, for there are examples where we have acted in time, and these provide valuable lessons for the case at hand.

Factors That Make It Difficult to Act in Time

First, it is often difficult to act in situations with a great deal of *uncertainty*. Social psychologists teach us that people often tend toward optimism and avoidance. The deeper truth is that human beings are very bad at dealing with uncertainty, and the problem is not always that they underestimate danger. When the "D.C. sniper" was active in the Washington, D.C., area, almost no one was on the streets. I assume one's odds of dying in the city had hardly changed, but people completely changed their behavior. This shows how uncertainty can lead to bad decisions.

The first and obvious scenario where uncertainty is an action stopper is when one may not know for sure that there is a problem. In this case avoidance wins the day. Even if one is certain a problem exists, when there's uncertainty about the causes, acting in time is tough. And when there's uncertainty about the consequences and timing of a future catastrophe, it gets even tougher. This sort of uncertainty interacts with a political culture that is inevitably focused on the next election. Everyone knew that a hurricane would hit New Orleans at some point in time, but no one knew when. Any politician who had acted to strengthen the levees was unlikely to see the benefits during a term in office. He or she would only face the economic and political costs associated with doing the hard work.

Perhaps the worst form of uncertainty involves being unsure of what works. It is hard to ask a leader to tackle the current costs for implementing a policy that may not even solve the problem. This volume helps reduce our uncertainty about the problems, about the consequences, and about which solutions actually would be effective. As such, it should help nations to act more readily. Sadly, these are not the only barriers.

Cases where there are *concentrated and immediate costs* required to achieve *diffuse and distant benefits* are tremendously challenging. If it is very clear who is going to pay the cost and it is not all that clear who is going to gain how much from acting, one has an energized set of opponents and a weak group of proponents. This is doubly amplified if the people who pay the costs are powerful and rich, and the people who benefit are poor, distant (or unborn), and not so powerful. The situation is further compounded when the whole point of acting is to prevent something from happening at all. No one even notices that disaster never struck, so only the up-front costs are visible. In the case of climate change, large carbon producers are easily identified. The swelling populations in Africa, whose children may go unfed if climate change dampens the agricultural potential of the continent, are not likely to be heard or politically potent even if they are engaged.

Third, where there are *competing equity claims*, action is muted. Most natural scientists tend to look for the best technical solution for achieving the desired result. Economists tend to want to find the most economically efficient mechanism. But the rest of humanity cares more about who is to blame than figuring out the most effective solution. Most people believe that whoever is to blame should have to pay to fix it. When a problem presents valid but competing views about equity and blame, then finger pointing replaces action. Who is to blame? Is it the producers or the consumers? Is it the developed world that's been using so much energy, or the developing world that is expanding its energy use so much? This factor is also magnified in a political environment. It is nearly impossible to imagine a candidate running for president of the United States arguing that it is essential to provide significant subsidies to China to get the Chinese to adopt carbon capture technology for their new coal-fired power plants. No one on the political scene talks that way, yet from what I can tell, almost every serious observer believes that the climate change solution is inevitably going to involve financial transfers to China.

Here is a factor that may also surprise: the presence of *too many solutions* also makes it harder to act in time. I am not talking about the case noted earlier where the uncertainty is about what will work. Instead, this is the case where many things could work, but just one needs to be chosen. Ironically, such choices often do not help. If there is only one thing that can be done, the body politic or the decisionmakers just have to decide whether the action is worth it. But often there are multiple options, such as whether to use carbon capture, renewable, or nuclear energy technologies, whether to have a carbon tax or tradable emission permits, whether to reduce carbon dioxide emissions or shoot sulfur or reflective particles into the atmosphere to deflect sunlight and cool the earth. Each option typically has very different fiscal impacts and calls forth alternative equity conceptions. Leaders and their supporters can claim that they really want to do something but that others are pushing a solution that is unfair and wrong-headed. They oppose the particular solution, not the right solution. The result is stalemate.

Finally and quite centrally, when solutions cross *sectors*, such as business, government, and civil society, and, even more important, when they cross *jurisdictions*, generating coordinated and effective action can be almost impossible. The opportunities for blame shifting, cost avoidance, and free riding are overwhelming. And energy and environmental challenges certainly cross sectors and jurisdictions as much as any modern issue. A solution will require that business, government, and civil society work effectively together, and that many different nations and localities voluntarily enforce what is sure to be a complex regime. Hurricane Katrina in New Orleans was hard enough, and it just involved regional, federal, state, and local leaders.

In combination, these factors make the challenges related to energy and climate change seem nearly insurmountable, especially when we are not even able to act in time to address the impending Social Security funding problem, which has almost none of these adverse elements. Regarding Social Security funding, there is almost no uncertainty, only one jurisdiction (the federal government), and equity is not much of an issue. Still, the energy issue is so monumental that there is a source for hope, for there are some occasions when we have acted in time in the past, and one can learn from those episodes.

Factors That Facilitate Acting in Time

In cases where timely steps have been taken, someone or something has managed to make the problem *vivid* in the present, even though the bigger consequences are in the future. Vividness is crucial to forcing action. The sheer scale of the potential impacts of climate change actually helps the situation. More important, the visually compelling aspect of early indications helps a great deal. Polar bears adrift on ice sheets or melting glaciers or dramatic weather events may or may not be traceable to climate change, but if the public believes they are, it is more inclined to call for answers. The goal must be to make the immediate indicators so compelling that people are willing to make significant sacrifices.

The second interesting feature about solutions is that they almost always involve *strange bedfellows*, that is, unusual coalitions are formed, in part out of self-interest. When companies decide that a change in regime will help them reap larger profits, they become advocates for movement. If market or regulatory uncertainty is making it hard to plan, produce, or sell products, business may get on the bandwagon. Most effective are occasions when businesses and groups determine that they can benefit from a new solution—for instance, their new innovations in alternative energy will become far more valuable if carbon becomes expensive. And for some energy producers, the uncertainty about what the price of energy will be in the future may be more damaging than a certain price rise.

Third, nations need to find some way to create *accountability* for inaction. Cross-jurisdictional barriers can be overcome if others notice and if the offending entity is evident to all. One must ensure that people really are behaving in the ways that they claim to be. Without this accountability, people will avoid action because they fear free riders.

And last, but probably most obvious and also most important of all, *leadership* is absolutely essential. When we do take wise and precautionary steps and act in time, a leader has typically come forward and chosen to take on the challenge to convince people there is both a problem and a solution. It now appears that leaders in several nations are coming forward. The obvious question is whether President Barack Obama will be such a visionary leader.

In the pages that follow, the reader will learn of specific issues, options, and opportunities. He or she will learn the state of current technology and the pressures for action and inaction. One can only hope that as a result, we all are more likely to act in time.

Acknowledgments

This book is the product of a great many people. Let me begin by recognizing the extraordinary contributions of the authors, my esteemed colleagues and good friends Laura Diaz Anadon, Max Bazerman, Bill Hogan, John Holdren, Henry Lee, and Dan Schrag, all of whom enthusiastically agreed to collaborate on this book. David Ellwood also deserves special credit for formulating the "acting in time" concept and inspiring all of us to exploit this useful framework. I thank all my colleagues in the Energy Technology Innovation Policy research group at the Belfer Center for Science and International Affairs, Harvard Kennedy School. Parts of this book draw heavily from the research of the group, other publications from which can be found at www.energytechnologypolicy.org. The meticulous editing and proofreading of Greg Durham and Patricia McLaughlin is greatly appreciated. Three research assistants, Benjamin Ganzfried, Brendan Luecke, and Devpriya Misra, were also very helpful in many aspects of the work.

The chapters were originally presented as papers at the Conference on Acting in Time on Energy Policy, held at the Harvard Kennedy School in September 2008; they were then revised by the authors in light of extensive insightful comments by the panelists and participants at the conference and by anonymous peer reviewers. The conference was cochaired by Bill Hogan and myself under the auspices of the Consortium for Energy Policy Research at Harvard. A short video from the conference is available at the website listed above. Experts from across the United States from the private sector, government, nongovernmental organizations, the labor community, and academia provided discerning commentary while serving on the conference panels, and I and the other organizers are grateful to all of them: Jane "Xan" Alexander (Clean Air–Cool Planet), Ashton Carter (Harvard), Ralph Cavanagh (Natural Resources Defense Council), John Deutch (MIT), Jeffrey Frankel (Harvard), Ben Heineman (Harvard), Gardiner Hill (BP Alternative Energy), Betsy Moler (Exelon), Ernie Moniz (MIT), Venkatesh Narayanamurti (Harvard), Don Paul (Energy and Technology Strategies), Paul Portney (University of Arizona), Dan Reicher (Google.org), Alan Reuther (United Auto Workers), Jim Rogers (Duke Energy), Phil Sharp (Resources for the Future), Todd Stern (Wilmer-Hale), Sue Tierney (Analysis Group), Jim Woolsey (VantagePoint Venture Partners), David Victor (Stanford), and Zou Ji (Renmin University).

I also express my gratitude to Floyd DesChamps and Jason Grumet, who as dinner speakers eloquently described the energy policy positions of the two presidential candidates. My thanks go as well to the other Harvard faculty program committee members not already mentioned: Graham Allison, George Baker, James Hammett, Jim McCarthy, Michael McElroy, Dale Jorgenson, Forest Reinhardt, Jack Spangler, Robert Stavins, and Richard Vietor. The conference would not have been possible without the diligent supporting efforts of Trudi Bostian, Greg Durham, Sam Milton, Jo-Ann Mahoney, Robert Stowe, and Amanda Swanson.

The Harvard contributors wish to express their gratitude to the following entities for their support: Belfer Center for Science and International Affairs at the Harvard Kennedy School, Bank of America, BP Alternative Energy Holdings, BP International, Henry Breck, the Doris Duke Charitable Foundation, the Energy Foundation, the William and Flora Hewlett Foundation, the David and Lucille Packard Foundation, Shell Oil, the U.S. Environmental Protection Agency, and Raymond Plank and the Apache Corporation.

Finally, I thank Kevin, Theodore, and Estelle Gallagher for inspiring and sustaining me in countless ways, large and small. I also wish to express my deep appreciation to the editors and staff at Brookings Institution Press for their early and consistent commitment to this project.

one
Acting in Time on Energy Policy
Kelly Sims Gallagher

This book clarifies the urgent priorities for U.S. energy policy at the dawn of the Obama administration and recommends specific steps that the U.S. government should take to address the numerous energy-related challenges facing the United States. Government must play a prominent role to ensure that adequate supplies of various forms of energy are available to enable and sustain U.S. economic growth, boost the competitiveness of U.S. firms in the global energy marketplace, counter the extreme volatility in oil prices of the past few years, limit the political and economic vulnerabilities associated with dependence on oil and natural gas, adequately and cost-effectively address the global climate change threat, and develop, acquire, and deploy advanced, clean, and efficient energy technologies to meet all of the above challenges.

The book's title—*Acting in Time*—refers to the persistent problem in U.S. energy policy that typically just enough is done to satisfy the short-term political imperatives, but not enough is done to actually solve the underlying problems themselves. As a result, many of the fundamental economic, environmental, and security-related challenges arising from patterns of U.S. energy production and consumption have become more intractable. Some now approach a point of crisis.

The United States is hugely influential in global energy markets, and in turn, international energy resources, supplies, and prices are central to the economic and environmental health of the United States. As the world's biggest economy, the United States is the largest national energy consumer, largest electricity producer, largest oil consumer, largest oil importer (twice as large as the second-largest oil importer, Japan), largest refiner of crude oil (in terms of crude distillation capacity), largest natural gas importer, largest producer and consumer of nuclear power (it has almost twice as much installed capacity as France), and the largest emitter of carbon dioxide (the dominant greenhouse gas [GHG]) on a per capita basis within the Organization for Economic Cooperation and Development, together with Australia.[1] The United States is the second-largest coal consumer, second-largest GHG emitter in aggregate terms, and fourth-largest producer of hydroelectric energy.[2] As for renewable energy, the United States is the largest ethanol producer and ranks third in terms of new renewable energy capacity investment and overall capacity.[3] Notably, U.S. wind installations in 2007 were not only the largest on record in the United States but were more than twice the previous U.S. record, set in 2006. Despite the widespread perception that the United States has fallen behind Europe in renewable energy development and deployment, no country in any single year has added the volume of wind capacity that was added to the U.S. electrical grid in 2007.[4]

Internationally, 61 percent of the world's proven oil reserves are located in the Middle East.[5] The global proven oil reserves controlled by nationally owned oil companies are estimated to range from 80 to 90 percent.[6] Most currently economically recoverable natural gas reserves are located outside the United States, with 67 percent located in the Middle East and Russia.[7] Rapidly growing industrializing countries are demanding increasingly more energy services. Between 2006 and 2008, for example, Asian oil demand grew on average 1.5 million barrels a day each year.[8] Primary energy demand in India and China is projected by the International Energy Agency (IEA) to

1. International Energy Agency (IEA, 2008).

2. Qatar, Kuwait, Bahrain, Australia (just slightly), Trinidad and Tobago, and the United Arab Emirates all have higher CO_2 emissions per capita (measured in tons CO_2 per capita) than the United States. U.S. per capita CO_2 emissions were 19 tons CO_2 per person in 2006 compared with a world average of 4.28 tons CO_2 per person. In terms of total national emissions, China is believed to have surpassed the United States in 2007, though its per capita emissions are 4.5 times smaller than U.S. per capita emissions (IEA 2008; British Petroleum [BP] 2008).

3. Renewable Energy Policy Network for the Twenty-First Century (2007).

4. Wiser and Bolinger (2008).

5. BP (2008).

6. Pirog (2007).

7. BP (2008).

8. U.S. Energy Information Administration (EIA), "International and United States Total Primary Energy Consumption, Energy Intensity, and Related-Data Tables" (www.eia.doe.gov/emeu/international/energyconsumption.html [December 2008]).

more than double by 2030 but only grow by 25 percent in the United States during the same time period.[9] Already about two-thirds of India's oil imports and 45 percent of China's oil imports come from the Middle East, and their dependence on that region is certain to grow.

With oil prices set in a global market, the degree of U.S. economic vulnerability is proportional to its *total* oil dependence, not just *import* dependence. And the United States is by far the largest oil consumer in the world. Consumers must pay the market price increase on every gallon or barrel consumed, not just on the imported barrels, although the degree of import dependence certainly affects who gets the revenue.[10] Lately oil imports have accounted for about one-third of the U.S. trade deficit.[11] The U.S. transportation sector accounts for the majority of U.S. oil consumption and the largest and fastest-growing fraction of U.S. greenhouse gas emissions, so it is a ripe target for policy attention.[12]

With the election of President Barack Obama, a window of opportunity to change U.S. energy policy has opened again. Each new presidential administration offers the possibility of change, a theme upon which President Obama himself vigorously campaigned. A cautionary tale can be told based on the history of promises made during presidential campaigns and subsequent energy policies enacted. Every presidential candidate since the oil shocks of the 1970s has campaigned on a theme of energy "independence," but every subsequent president left office with the United States more, not less, dependent on foreign oil. U.S. oil imports as a percentage of total oil consumption were 37 percent during the Nixon administration, and they rose to 66 percent during the Bush administration.[13] On the subject of climate change, the record is no better. The Clinton-Gore administration entered office on the heels of the United Nations Conference on Environment and Development in Rio where the UN Framework Convention on Climate Change had been negotiated. Vice President Al Gore had just published his famous book *Earth in the Balance* (1992), which stressed the importance of action to tackle climate change. But despite the Clinton administration's initial proposal for an energy usage tax ("BTU tax"), subsequent negotiating of the Kyoto Protocol, and development of a voluntary domestic climate change action plan, no enforceable policies were enacted during its eight-year term. The Bush

9. IEA (2008).
10. See chapter 4 in this volume.
11. Jackson (2007).
12. Gallagher and Collantes (2008).
13. National Commission on Energy Policy (NCEP, 2007).

administration declined to support mandatory policies to reduce GHG emissions, and U.S. carbon dioxide emissions grew another 4 percent between 2001 and 2007.[14] U.S. government investments in energy research, development, and demonstration are approximately half the level they were thirty years ago in constant dollars. Japan now exceeds the United States in total government investments in energy technology innovation as well as in government investments as a percentage of GDP.

During his campaign, President Obama made many noteworthy and important commitments, namely, that he would

—invest approximately $15 billion a year for ten years in cleaner energy, thereby creating approximately 5 million associated jobs;

—reduce oil imports by volumes equal to the imports from the Middle East and Venezuela within ten years;

—modernize the national electricity grid;

—commit to reduce carbon emissions to 1990 levels by 2020 and to effect an additional 80 percent reduction by 2050 via a market-based cap-and-trade system with a full auction, as well as to reengage in international climate negotiations;

—give every family a $1,000 energy rebate and pay for it from oil company profits;

—provide $4 billion in loans and tax credits to American auto plants and manufacturers so that they can retool factories and build fuel-efficient cars;

—put 1 million 150-mpg, plug-in hybrids on U.S. roads within six years and give consumers a $7,000 tax credit to buy fuel-efficient cars; and

—ensure that 10 percent of U.S. electricity comes from renewable sources by 2012 and 25 percent by 2025, as well as extend the production tax credit for renewable energy for five years.

Given the severe financial and economic challenges facing the new president, it is natural to wonder if energy policy should be a top priority at the beginning of the new administration. If this book does nothing else, it makes the case that the United States cannot afford to wait any longer to enact long-term policies for climate change, carbon capture and storage (CCS), electricity structure reform and infrastructure investment, oil security, and energy-technology innovation. Tempting as it may be to defer policymaking in the energy and climate domains, not taking prompt action early in the Obama administration will constitute a failure to act in time. In fact, acting early is clearly in the longer-term financial interest of the United States.

14. EIA (2008).

Some, including President Obama, have argued that the financial crisis presents an opportunity for investing in a "green" recovery. One can thus readily imagine that President Obama's commitment to invest $15 billion a year in the development and deployment of cleaner energy technologies will be acted upon because such investments are likely to create new jobs and enhance the industrial competitiveness of U.S. firms, which are increasingly at risk of arriving too late to the global marketplace for advanced energy technologies. Eight of the ten top wind manufacturers are European, none of the top five global producers of photovoltaic cells is based in the United States, and two Japanese producers supply 85 percent of the world's market for hybrid electric vehicle batteries.[15] Numerous suggestions for how to best deploy and utilize these proposed funds are outlined in chapter 5, on energy-technology innovation. One likely source of these funds could come from the passage and implementation of domestic climate change legislation, which would generate revenues if carbon dioxide emissions are taxed or if permits are auctioned rather than given away.

There will be tensions as the Obama administration endeavors to tackle all the energy challenges. One consistent theme in this book, for example, is that higher energy prices would help achieve all the policy objectives in the longer term—improved oil security, lower GHG emissions, more efficient operation of the electricity system, more incentives for private sector innovation in energy technologies, and more incentives for consumers to purchase cleaner and more energy-efficient products. But, of course, higher energy prices have not been deemed politically palatable in the past, and there is the legitimate concern that they can unfairly burden low-income Americans. To address the disproportionate impacts on low-income families, rebates could be provided or social welfare programs enhanced. Policymakers need to break out of the trap of reinforcing energy problems by suppressing energy prices, and there is evidence to suggest this is doable. For example, there is substantially more support for carbon taxes, which increase energy prices, when they are presented in the context of other public finance choices, such as reductions in income taxes. Polling from 2006 shows that support for the same carbon tax level tripled and opposition fell by two-thirds when a large carbon tax was paired with a similarly large income tax cut.[16]

Although there were many possible focuses within the realm of energy policy for this book, we chose to concentrate on six topics in particular: climate

15. See Anadon and Holdren, chapter 5 in this volume.
16. Ansolabehere (2006).

change policy, CCS policy, oil security policy, energy technology innovation policy, electricity market structure and infrastructure policy, and barriers to acting in time on energy policy and strategies for overcoming them. Each of these topics is the focus of one full chapter, and policy recommendations are provided at the conclusion of each chapter. These topics were chosen because they are arguably the six most important and urgent focuses for the Obama administration. Each is considered in turn below.

More than ten years have passed since the Kyoto Protocol was adopted, and since the turn of the century, 26 billion tons of CO_2 a year have been emitted from the burning of fossil fuels (with another 4–8 billion tons emitted from land use change and deforestation) on average. Given the large amounts of GHGs that are already in the atmosphere, some climate change is now virtually inevitable. Scientists estimate that about one-third of mid-twenty-first-century warming is already committed given the amount of GHGs already in the atmosphere, but the other two-thirds of projected warming is strongly dependent on how much more is discharged into the atmosphere during the next two decades. No matter which ultimate GHG concentration target is chosen, global GHG emissions reductions are necessary and inevitable, but the hard questions are when each country should begin, and how fast it should reduce its emissions. In chapter 2 on climate change, I argue that the world, and the United States specifically, must first establish a long-term GHG concentration goal. Once there is agreement on this goal, the required avoided emissions can be determined, and a long-term GHG "emissions budget" can be created. An emissions budget is no different from a financial budget in that it simply provides a quantitative limit on emissions (that is, spending) for a given time period.

Procrastination—or failing to act in time—results in much faster required rates of emissions reductions if a certain concentration target is to be met. The faster the pace of required emissions reductions, the more difficult and expensive it becomes to live within the budget. Too much delay and the budget is blown, meaning that our children and grandchildren must cope with much larger magnitudes of climate disruption. Therefore, the three most important steps the U.S. government must take with respect to climate change are to, first, set the long-term goal—recognizing that this goal may need to be revised in light of new scientific information—and thereby create a national emissions budget; second, place an initial price on U.S. greenhouse gas emissions, either through a cap-and-trade mechanism or a tax; and third, reengage internationally, especially with China—the world's largest and fastest-growing emitter—to devise an international solution to the climate change challenge.

With regard to both climate change and energy technology innovation policy, chapter 3 is purely devoted to CCS. Why focus on this technology rather than on technologies for renewable energy, nuclear energy, or energy efficiency? The answer is that there will be a desire to rely increasingly on the vast U.S. coal reserves, both because of their abundance and relative cheapness—if one excludes all the health and environmental costs associated with coal.[17] The United States has a huge number of concentrated point sources of CO_2. If a way can be found to economically capture and store the carbon from those factories and plants, they will not have to be retired prematurely at great economic cost, which is unlikely to happen anyway. There are more than 500 power plants that emit more than 1 million tons of CO_2 in the United States alone. In chapter 3 Daniel Schrag outlines the scientific and technical challenges for large-scale carbon capture (both precombustion and postcombustion) and carbon storage. Even though the technical challenges are significant, CO_2 is already economically captured, transported, and used for enhanced oil and gas recovery all over the world. Costs for the first set of capture and storage facilities appear to be high (too expensive to be motivated by relatively low carbon prices), but Schrag notes that these costs will come down through further research, development, and demonstration. He recommends that the U.S. government provide federal subsidies for ten to twenty commercial-scale CCS projects and argues that these demonstration projects should use different capture technologies, employ different strategies for geological storage, and be spread across different regions of the United States to have the biggest impact, both on knowledge gained and public perception. Schrag also recommends that new federal laws and regulatory policies should be created so that developers and operators of power plants and CO_2 storage facilities understand their liability, and know which environmental regulations will apply to CCS projects. Third, he contends that the federal government should encourage state and local governments to accelerate permitting processes for CCS projects. Finally, he argues that the long-term goal should be the adoption of CCS for all large stationary sources. Acting in time on CCS thus requires that a major research, development, and demonstration program for CCS be implemented immediately so that knowledge can be acquired about the viability of this technology during the next five to ten years, and so that costs can be brought down to a reasonable level by the time deep cuts in CO_2 will be required in the United States to meet our long-term climate goals.

17. As Schrag notes in chapter 3, although coal is the major motivation for the development of CCS, CCS need not only apply to coal: any point source of CO_2 can be sequestered, including CO_2 from combusted biomass, which would result in negative emissions.

In terms of oil security, although President Obama campaigned on a theme of energy independence, Henry Lee points out in chapter 4 that the United States is not its own market but rather part of an international market. Oil independence, therefore, is neither affordable nor desirable. Lee defines the oil security problem as encompassing four concerns: short-term economic dislocations from sudden increases in oil prices, long-term supply inadequacies, a foreign policy overly constrained by oil considerations, and environmental threats, specifically global climate change. He highlights the problems associated with the volatility in oil prices and notes that low prices could easily be followed by an era of high oil prices again. To address these concerns, the growth in world oil consumption and GHG emissions needs to be reduced. To accomplish these goals, Lee suggests that the United States needs to place a price on both imported oil and carbon, either through taxes or a cap-and-trade program. By increasing prices, the government sends a strong signal to consumers to use less oil and emit fewer grams of carbon, and it also sends a powerful signal to entrepreneurs and investors who are striving to develop substitutes for imported oil. Lee suggests that Congress consider a variable tax that would be triggered when oil prices reach a certain threshold—for example, $90 per barrel—so if oil prices slipped below $90 to $80 per barrel, a $10 tax would be imposed. If the price later rose above $90, the tax would disappear. Politically, this proposal seems unlikely during an era of low oil prices, but the government should be prepared to take advantage of this opportunity when oil prices rise again. There are other policy options as well, but a key point is that international cooperation is crucial because a coordinated effort that reduces oil consumption in all the major oil-importing countries will be the most effective way to improve oil security. There are few short-term fixes to the oil security problem, so acting in time on oil security requires devising strategic and consistent domestic and foreign policies that will improve U.S. oil security over the longer term, hopefully before the next crisis hits.

Policy for energy technology innovation is important because innovation can both reduce the costs of energy technologies today as well as improve the menu of options for the future. In chapter 5 Laura Diaz Anadon and John Holdren contend that current U.S. public and private energy research, development, and demonstration expenditures are small in relation to the economic, environmental, and security stakes. To move cleaner and more efficient energy technologies from the laboratory into the marketplace, "market-pull" policies are necessary complements to "technology-push" policies, and there should be much greater coordination between the push and pull policies than has existed until now. Acting in time in the energy innovation domain is necessary if the United States is to maintain its role as a flourishing and competitive economy, give the world a chance to prevent a climate disruption crisis, and minimize the chances of fossil fuel or nuclear energy related international conflicts. To illustrate how important investments in innovation could be for the U.S. economy, consider that growing demand for energy and low-carbon technologies is creating a large market for advanced energy technology suppliers estimated at approximately $600 billion a year. As Anadon and Holdren point out, whether or not U.S. firms will capture a large fraction of this market depends on the right policy conditions and incentives.

A key challenge for electricity market design and regulation is to support efficient investment in infrastructure. In the coming years, if the United States is to have adequate electricity supply or be able to increase the fraction of renewable energy in the electricity mix, for example, a workable regulatory and market framework is essential. William Hogan explains in chapter 6 that initiatives to improve energy security, meet growing demand, or address climate change and transform the structure of energy systems all anticipate major electricity infrastructure investment. Without the necessary infrastructure investment, energy policy cannot take effect, and without sound policy, the right infrastructure will not appear—a classic chicken-and-egg problem. Acting in time thus requires that policies are put into place now to support efficient investment in infrastructure so that all the other desirable energy policies can be implemented. As discussed at length in Hogan's chapter, improved scarcity pricing and a hybrid framework for transmission investment are two workable solutions that seem necessary to meet the needs for a long-term approach to infrastructure investment.

The likely barriers to the policies prescribed in this book and the strategies for overcoming them are identified in the final chapter by Max Bazerman. People often fail to act in time to prevent foreseeable catastrophes—"predictable surprises"—and we see several such surprises looming with respect to U.S. economic competitiveness, oil security, and global climate change. Bazerman argues that enacting wise legislation to act in time to solve energy problems requires surmounting cognitive, organizational, and political barriers to change. To take the example of the organizational barriers, it is clear that the U.S. government is not presently structured in a way that would allow it to forcefully confront our current energy challenges. Moreover, government employees are often not trained in the methods needed to implement smart energy policies. While there is a Department of Energy, civilian energy is a very small part of the department's budget and activities. No single agency or entity is in charge of collecting information on climate change or energy security from all the departments, analyzing that information, and transforming it into effective policy. Institutions are composed of the laws, rules, protocols, standard operating procedures, and accepted norms that guide organizational action, and members of these institutions come to behave by force of habit. This makes many bureaucratic government departments resistant to voluntary information sharing and regulatory flexibility.

The Obama administration and new Congress, Bazerman argues, must anticipate and address aspects of government organizations that will prevent successful implementation of new ideas aimed at acting in time to solve energy problems. Bazerman suggests five ways to improve the odds that the policies recommended in this book will succeed. First, policymakers should identify policies that make wise trade-offs across issues and then, second, communicate to the public that decisions are being made to maximize the overall benefits to U.S. society rather than to special interest groups. Third, the new administration should devise energy policies that make sense even if climate change were less of a problem than best current estimates suggest. Fourth, the administration should identify a series of small changes or "nudges" that significantly influence the behaviors of individuals and organizations in a positive direction. Finally, when discounting of the future creates an insurmountable barrier to the implementation of wise policies, Bazerman suggests considering a mild delay in implementation.

It should be noted that the energy policy challenges and priorities for the United States are not the same as for many other countries. The U.S. context is unique, and this book focuses primarily on the problems and opportunities that exist in the United States. Other major oil importers, like Japan, Europe, and China, share many of the same oil security problems. Other major coal consumers, including China and India, should be similarly motivated to consider CCS. The United States and China, as the top two GHG emitters on an aggregate basis, will need to think about their carbon mitigation strategy during the coming decades, just as Europe and Japan have already done. Nearly all countries have an incentive to develop strategies for energy technology innovation to address the distinctive needs of their own countries. The challenges with respect to the U.S. electricity grid and electricity markets are perhaps the most specially American given the elaborate regulatory structure at both the federal and state levels.

Finally, although assigned to no special chapter in this book, energy efficiency permeates the basic themes and should be considered a top priority for the U.S. government. The United States lags far behind most other industrialized countries in addressing the overall energy intensity of its economy, ranking thirteenth among the industrialized countries and fifty-fifth over-all.[18] Greater energy efficiency offers leverage against all of the major economic, security, and environmental problems faced by the United States. Greater energy efficiency improves oil security and reduces GHG emissions, the need for greater power supply capacity, and pressure on the electrical grid. It reduces the amount of money being spent on oil and gas imports, and it can improve the productivity of U.S. firms. In fact, a worthy goal would be for the United States to become the most energy-efficient economy in the world.

References

Ansolabehere, Stephen. 2006. "U.S. Public Attitudes on Swapping Carbon Taxes for Income Taxes." Harvard University, Department of Government.

British Petroleum (BP). 2008. *BP Statistical Review of World Energy.* London.

Gallagher, Kelly Sims, and Gustavo Collantes. 2008. "Analysis of Policies to Reduce Oil Consumption and Greenhouse Gas Emissions from the Transportation Sector." Discussion Paper 2008-06. Cambridge, Mass.: Belfer Center for Science and International Affairs (June).

Gore, Al. 1992. *Earth in the Balance: Ecology and the Human Spirit.* New York: Houghton Mifflin.

International Energy Agency (IEA). 2008. *Key World Energy Statistics.* Paris.

Jackson, James K. 2007. "U.S. Trade Deficit and the Impact of Rising Oil Prices." Report RS22204. Congressional Research Service, Library of Congress.

National Commission on Energy Policy. 2007. *Collaborative Development of Petroleum Sector Performance Indicators (PSI).* Washington.

Pirog, Robert. 2007. "The Role of National Oil Companies in the International Oil Market." Report RL34137. Congressional Research Service, Library of Congress.

Renewable Energy Policy Network for the Twenty-First Century. 2007. *Renewables 2007 Global Status Report.* Paris: REN21 Secretariat, and Washington: Worldwatch Institute.

U.S. Energy Information Administration (EIA). 2008. "U.S. CO_2 Emissions from Energy Sources." (May).

Wiser, Ryan, and Mark Bolinger. 2008. *Annual Report on U.S. Wind Installation, Cost, and Performance Trends 2007.* DOE/GO-102008-2590DOE/GO-102008-2590. U.S. Department of Energy, Energy Efficiency and Renewable Energy (May).

18. See EIA, "International and United States Total Primary Energy Consumption." Energy intensity is defined as total primary energy consumption per dollar of gross domestic product, using market exchange rates. If energy intensity is calculated using purchasing power parity estimates rather than market exchange rates, the United States ranks 144th overall.

two
Acting in Time on Climate Change

Kelly Sims Gallagher

In 2008 Arctic sea ice melted at a record rate, and 1,100 daily precipitation records were broken throughout the Midwest in the month of June alone. In 2007 global temperatures were the fifth warmest since 1880. Six of the ten warmest years on record for the United States have occurred since 1998. These unusual events, and others, have created renewed concern about the threat of global climate disruption. The United States urgently needs a coherent climate policy.

Despite the entry into force of the Kyoto Protocol in 2005 and innumerable efforts around the world to reduce greenhouse gas (GHG) emissions, global emissions have continued to grow steadily. Carbon dioxide emissions from energy and cement production grew 34 percent between 1990 and 2007 globally. U.S. GHG emissions grew 18 percent during that same time period. Just since the turn of the century, global CO_2 emissions have grown 16 percent. While the United States has long been the world's biggest GHG emitter, China's aggregate emissions surpassed U.S. total emissions in 2006. As of 2007, the two countries together accounted for 46 percent of global CO_2 emissions. Since neither of the world's biggest emitters have enacted clear national policies designed to reduce GHG emissions, and since global emissions are growing much faster than previously expected (about 3 percent a year so far this century), the question of whether governments are failing to act in time to address the threat of climate disruption must be asked.

This chapter explores a number of related questions: How much time do we have to act? How much climate change is virtually inevitable? What are appropriate emission pathways? What are the consequences of procrastination? And finally, what is the appropriate role for governments wishing to act in time to reduce the threat of climate change? In addition, the reality of current emissions and policy responses is explored in some detail for the two biggest emitters in the world: the United States and China.

How Much Time Do We Have to Act?

The first question to consider is how much time remains to act decisively enough to avoid catastrophic climate disruption. It is now clear that some climatic change has begun and that more is inevitable, but future damage could range from small to large depending on how soon global emissions start to decline, how much they are reduced, and by when. There is simply no definitive answer to the question of how much time there is to act because of the uncertainties related to the sensitivity of the global climate system, the inherent limitations of global climate models, and the possibilities for unpleasant surprises.[1] Still, the available information in the scientific literature can shed considerable light on the question.

Before the industrial revolution, the level of carbon dioxide in the atmosphere was 280 parts per million (ppm). Since that time, humankind's emissions of CO_2 have grown steadily, reaching 384 ppm in 2007.[2] Annual CO_2 emissions from burning fossil fuels since 2000 have averaged about 26 billion tons a year, and annual emissions from land use change and deforestation are estimated to be an additional 4–8 billion tons a year globally.

The increased concentrations of CO_2 and other anthropogenic greenhouse gases in the atmosphere have already caused changes in the composition of the atmosphere. Aside from CO_2, the dominant anthropogenic GHG, the other climate-altering gases are methane (CH_4), nitrous oxide (N_2O), chlorofluorocarbons (CFCs) and hydrochlorofluorocarbons (HCFCs), perfluorocarbons (PFCs), hydrofluorocarbons (HFCs), and sulfur hexafluoride (SF_6). As of 2005, these other GHGs currently had added about 50 ppm carbon dioxide equivalent (CO_2e) in terms of current concentrations of CO_2 in the atmosphere. The effects of particles (warming from some, cooling from others) added up to

1. The sensitivity of the climate system can be viewed as how hard the system can be pushed or stressed before it reacts unpredictably.
2. National Oceanic and Atmospheric Administration, "Trends in Atmospheric Carbon Dioxide—Global" (www.esrl.noaa.gov/gmd/ccgg/trends).

approximately a net negative 50 ppm CO_2e, so the effective concentration in 2005 was 380 ppm CO_2e. In response to the presence of these heat-trapping gases in the atmosphere, global average surface temperatures increased about 1.3°F (0.7°C ± 0.2°C) during the last century, though the temperature changes varied widely in different regions according to the 2007 Intergovernmental Panel on Climate Change (IPCC) report.[3] Significant increases in precipitation have been experienced over eastern North and South America, northern Europe, and northern and central Asia. Drying has been observed in the Sahel, the Mediterranean, southern Africa, and parts of southern Asia.

As a result of these changes in the climate, some impacts already can be observed. Sea levels rose at an average rate of 1.8 millimeters a year between 1961 and 2003, mainly due to thermal expansion of the oceans, but the rate of increase was much faster between 1993 and 2003—about 3 mm a year during that decade. Mountain glaciers and snow cover have declined in many places, reducing the availability of fresh water supplies. The Greenland ice sheet has sustained heavy losses during the summer, and satellite data since 1978 show that the extent of Arctic sea ice has shrunk by about 7 percent a decade during the summer months.[4] In early September 2008, Arctic sea ice was diminished to its second-lowest level on record—the lowest was in 2007.[5] Wildfires in the western United States have increased four-fold in the last thirty years due in part to increased temperatures and earlier spring snowmelts.[6] Finally, incidence of severe flooding is sharply up in North America, Africa, Europe, and Asia.[7] It appears that time already has run out to altogether prevent climate change. The next question to explore then is how much more climate change is virtually certain to occur.

How Much Climate Change Is Virtually Inevitable?

Given the large quantity of GHGs that humankind has already deposited in the atmosphere, how much climate change is now inevitable? Greenhouse gases, including carbon dioxide, methane, and nitrous oxide, last in the atmosphere for decades to centuries. Perfluorinated compounds including sulfur hexafluoride, PFC_{14}, and PFC_{218} last for 2,600–50,000 years.[8] If emis-

---

3. IPCC Working Group I (IPCC WG1 2007).
4. Chylek, Dubey, and Lesins (2006).
5. National Snow and Ice Data Center, "Arctic Sea Ice Now Second Lowest on Record" (nsidc. org/arcticseaicenews/2008/082608.html [September 2008]).
6. Westerling and others (2006).
7. Hassan, Scholes, and Ash (2005).

sions were quickly reduced enough to stabilize GHG concentrations, global average temperature would continue to rise mainly because the thermal inertia of the ocean introduces a lag to the warming of the climate system even after concentrations are stabilized.[9] These are the reasons that some amount of future climatic change is virtually inevitable.

There is no way to predict the climate future with precision, but sophisticated models have been developed to depict, in at least an approximate way, the temperature increases and associated changes in other climate variables that would be associated with different future GHG emission paths and associated concentrations. The IPCC has concluded that even if the atmosphere were instantaneously stabilized at its year 2000 composition, global temperatures would increase about another 1.1°F (0.6°C) on average during the next several decades.[10] This means that about one-third of mid-twenty-first-century warming projected under "business as usual" emissions trajectories is already committed; but the other two-thirds of the amount of projected warming are strongly dependent on the quantity of GHGs that is released into the atmosphere between now and then.[11] In the IPCC's "best-case" scenario of relatively low emissions growth during this early part of century (followed by a rapid decline in global emissions), an additional 3.1°F (1.8°C) warming by 2100 is projected.[12] In the IPCC's "worse-case" scenario of relatively steady growth in GHG emissions throughout the twenty-first century (followed by emissions reductions during the twenty-second century), a 7.2°F (4°C) warming by 2100 is projected.[13] These emission scenarios are shown in figure 2-1, where the best-case scenario is depicted as scenario I and the worst-case scenario as scenario VI.[14] A more recent analysis indicates that even if there is no growth in emissions or concentrations, the world may already be committed to a 3.6°F (2°C) rise in temperatures in the long term (over centuries) due to slow "feedbacks."[15] Feedback mechanisms are changes in the climate system that lead to additional or enhanced changes in the climate, or both. A classic example of a "positive" feedback is the warming of the oceans. Oceans currently absorb some of the extra CO_2 in the atmosphere, but as the oceans warm in response to higher

8. IPCC WG1 (2007, p. 212).
9. Ibid., p. 822.
10. These are global average surface temperatures. They will vary widely around the globe.
11. IPCC WG1 (2007, p. 749).
12. This is the IPCC's "B1" scenario. The best estimate is 1.8°C, but the likely range is 1.1–2.9°C.
13. This is the IPCC's "A1F1" scenario. The best estimate is 4°C, but the likely range is 2.4–6.4°C (IPCC WG1 2007, p. 13).
14. IPCC Working Group III (IPCC WG3 2007).
15. Hansen and others (2008).

Figure 2-1. World Carbon Dioxide Historical Emissions and Scenario Projections, 1940–2100 (left side), and Projected Equilibrium Global Average Temperature Increase above Preindustrial for a Range of GHG Concentration Stabilization Levels (right side)

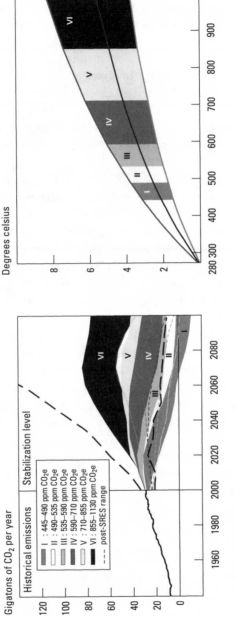

Source: IPCC (2007, figure 5.1).

a. SRES, Special Report on Emissions Scenarios.

surface temperatures caused by global warming, the capacity of the oceans to absorb CO_2 is decreased. Since the oceans cannot take up as much CO_2 as before, more CO_2 remains in the atmosphere, thereby causing additional warming. The vicious cycle thus continues.

Projecting climate disruption for the twenty-first century, the IPCC concluded that it was "virtually certain" that there would be fewer cold days and nights, and more frequent hot days and nights over most land areas. The IPCC also projected that it was "very likely" that there would be more frequent heat waves and heavy precipitation events in most regions, which obviously can lead to health problems in the first instance and flooding in the second.[16] Sea levels are expected to rise several meters over the next few centuries even with current concentrations of 384 ppm CO_2.

Of course, there could (and almost certainly will) be "surprises," some of which would actually not be so surprising since they already have been anticipated. Bazerman and Watkins call these "predictable surprises"—events that apparently take people by surprise, despite prior awareness of all the information necessary to anticipate the events and their consequences.[17] Several predictable surprises have already been identified by scientists, many of which may come in the form of climate feedbacks. Global climate models currently do not include all the potential feedbacks in the climate-carbon cycle, nor do they include the full effects of changes in ice sheet flow or the impacts of large releases of currently frozen methane clathrates. The longer the earth experiences high temperatures, the more likely that the soils and tundra will warm up and release currently frozen methane into the atmosphere, for example, which would further accelerate global warming or could even induce a so-called runaway greenhouse effect. Feedbacks that are now included in the models are increases in water vapor and decreased sea ice due to warmer temperatures, both of which would exacerbate global warming. Scientists have already warned that it is unlikely that the changes in the climate will be gradual and constant in the same way that emissions of GHGs have been gradually increasing during the past two hundred years. Abrupt climate changes occur when the climate system is forced to cross some threshold, "triggering a transition to a new state at a rate determined by the climate system itself and faster than the cause."[18]

One example of an abrupt and unforeseen environmental change was the formation of the stratospheric ozone hole, which had not been predicted by

16. IPCC WG1 (2007).
17. Bazerman and Watkins (2004).
18. See Committee on Abrupt Climate Change (2002).

scientists but was reported in 1985.[19] Until the appearance of the ozone hole, there had been a steady and relatively predictable decline in the total amount of ozone in the stratosphere due to increased concentrations of CFCs. At a certain threshold, however, the ozone losses were unexpectedly accelerated by the formation of polar stratospheric clouds during the cold Antarctic winter; these act as reservoirs of chlorine gas that depletes the atmospheric layer of ozone during the Antarctic spring (September to December), generating an ozone "hole." This example illustrates that the precise climatic consequences of different levels of GHG concentrations in the atmosphere may prove hard to predict. Scientists have developed a wide range of scenarios based on different climate models, but all of them currently assume that there will be no abrupt climate changes.

The IPCC's Fourth Assessment Report provides a best estimate for the increase in temperature in response to CO_2 concentrations of 550 ppm (a doubling of preindustrial levels) of 3.6–8.1°F (2–4.5°C) during the twenty-first century, with a most likely value of about 5.4°F (3°C).[20] More recently, Hansen and others argue that over a longer term, a 550 ppm concentration could eventually result in a much warmer 10.8°F (6°C) global average temperature increase due to very slow feedback processes such as ice sheet disintegration, slow vegetation migration, and GHG releases from soils, deep ocean sediments, and tundra that are not currently incorporated into the main climate models.[21] In other words, the shorter the duration that the earth "overshoots" and experiences high GHG concentrations, the more likely it is that climatic disruptions will be closer to those predicted by climate models available today. Hansen and colleagues recommend a CO_2 concentration target of no more than 350 ppm, well below today's level of 384 ppm, if dangerous climatic changes such as loss of fresh water from mountain glaciers, large-scale sea level rise, and destabilization of Arctic sea ice cover are to be avoided.[22]

Creating an Emissions Budget

No matter which emissions concentration target is chosen, GHG emissions reductions are necessary—but when to begin and how fast to reduce emissions? Once there is agreement on the appropriate GHG concentration target,

19. Cagin and Dray (1993).
20. IPCC WG1 (2007, p. 799).
21. Hansen and others (2008).
22. Ibid.

the required emissions reductions can be determined. In other words, a "GHG emissions budget" can be created. An emissions budget is no different from a financial budget in that it simply provides a quantitative limit on emissions (that is, spending) within a given time period.

Researchers at the Tyndall Center for Climate Change Research and Sussex Energy Group have developed a carbon budgeting approach that is instructive. First, they identify a target emissions concentration. The Tyndall UK scenario for 450 ppm, for example, resulted in the United Kingdom having a cumulative carbon-equivalent budget of 4.6 gigatons of carbon (GtC) to emit by 2050. Their analysis demonstrates that in order to have a 30 percent chance of not exceeding the 2°C temperature change threshold (which they associate with the 450 ppm concentration level), the United Kingdom must reduce its emissions 90 percent by 2050. They conclude the United Kingdom's "carbon bank balance" is falling quickly, with some 17 percent of the fifty-year budget already spent in the first six years of the twenty-first century. In other words, further procrastination could easily result in the United Kingdom exceeding its cumulative emissions budget. The sooner emissions reductions occur, the more likely the United Kingdom will be able to stay within budget.[23] A similar study for China has concluded that China's budget through 2050 would have to range between 70 GtC and 111 GtC, depending on whether a per capita or emissions intensity approach were used to establish the target.[24] The study's authors conclude that if Chinese emissions do not begin to decline in absolute terms by 2030, the risk of overshooting the budget rises dramatically, but they also conclude that earlier turning points seem very challenging to envision given the rapid rate of growth in carbon dioxide emissions there during the past decade.

In a 2008 analysis of global emissions pathways for a stabilization target of 450 parts per million by volume (ppmv) CO_2e, Anderson and Bows reach some daunting conclusions about the global emissions budget for the twenty-first century.[25] Assuming that stabilization at 450 ppmv CO_2e in fact offers a 46 percent chance of not exceeding 3.6°F (2°C) global average temperature increase, they conclude that *global* energy-related emissions would have to peak by 2015 and rapidly decline at a pace of 6–8 percent a year between 2020 and 2040, and then the world would have to fully decarbonize soon after 2050.[26] The authors further conclude:

23. Bows and others (2006); Bows and Anderson (2007).
24. Wang and Watson (2008).
25. Anderson and Bows (2008).
26. See Meinshausen (2006) on full decarbonization.

—If emissions peak in 2015, stabilization at *450 ppmv* CO_2e requires subsequent annual reductions of 4 percent in CO_2e and 6.5 percent in energy and process emissions.

—If emissions peak in 2020, stabilization at *550 ppmv* CO_2e requires subsequent annual reductions of 6 percent in CO_2e and 9 percent in energy and process emissions.

—If emissions peak in 2020, stabilization at *650 ppmv* CO_2e requires subsequent annual reductions of 3 percent in CO_2e and 3.5 percent in energy and process emissions.[27]

Clearly, for a given stabilization target, the more procrastination, the faster the pace of emissions reductions must be. As the pace of required emissions reductions increases, the difficulty associated with meeting the chosen target also increases.

Given that there is uncertainty about the sensitivity of the climate, it is essentially impossible for governments to be certain that a given stabilization target will result in an avoidance of catastrophic climate change. As the IPCC recently noted:

> The choice of short-term abatement rate (and adaptation strategies) involves balancing the economic risks of rapid abatement now and the reshaping of the capital stock that could later be proven unnecessary, against the corresponding risks of delay. Delay may entail more drastic adaptation measures and more rapid emissions reductions later to avoid serious damages, thus necessitating premature retirement of future capital stock or taking the risk of losing the option of reaching a certain target altogether.[28]

Thus governments face the difficult task of balancing the upfront costs of mitigation and adaptation against the risks of climate change itself as well as the risks of having to rapidly reduce emissions in a way that could prove highly costly. There are only three real options: first, mitigation (taking steps to reduce the pace and magnitude of the climatic changes being caused by GHG emissions); second, adaptation (taking steps to reduce the adverse impacts of the changes that occur); and third, suffering from the impacts not averted by either mitigation or adaptation.[29] The more emissions can be reduced in the near term, the more likely suffering can be avoided. No matter how much we mitigate, however, there will be some need to adapt to the

---

27. Anderson and Bows (2008, p. 3879).
28. IPCC WG3 (2007, p. 233).
29. Holdren (2008).

inevitable climatic changes. How much effort will need to be expended on adaptation and alleviation of suffering will depend upon how well we live within the emissions budgets.

In summary, by connecting recent emission trends with global emissions budgets for concentration targets, two main insights can be derived about failing to act in time. First, delay results in much faster required rates of emission reduction if a certain concentration target is to be met. Second, too much delay simply blows the budget and results in a much higher concentration level. Anderson and Bows are pessimistic: "Given the reluctance, at virtually all levels, to openly engage with the unprecedented scale of both current emissions and their associated growth rates, even an optimistic interpretation of the current framing of climate change implies that stabilization much below 650 ppmv CO_2e is improbable."[30] A 650 ppmv CO_2e stabilization level is associated with a best-estimate guess of a global average temperature change of 6°F (3.6°C), rapid ice sheet disintegration, alteration of ocean circulation, and substantial sea level rise.[31]

Potential Costs of Procrastination

There is a large literature about the costs and benefits related to climate change and emissions reductions, and this chapter does not aim to provide a comprehensive discussion of this literature.[32] Simply, some have argued that the possible economic damages from climate change are so large that they vastly outweigh the costs of mitigation. Others have argued that depending on the choice of one's discount rate, one's confidence in the current scientific understanding of the problem, and one's optimism about future technological developments that may make the problem more tractable, it is not so clear that the benefits of mitigating the threat of climate change exceed the costs of mitigation today. Nordhaus is very clear on this point:

> Climate change is a complex phenomenon, subject to great uncertainties, with changes in our knowledge occurring virtually daily. Climate change is unlikely to be catastrophic in the near term, but it has the potential for very serious damages in the long run. There are big economic stakes in designing efficient approaches to slow global warming

---

30. Anderson and Bows (2008, p. 3380).

31. IPCC WG1 (2007, p. 66).

32. See, for example, Pearce and others (1996); Stavins (1999); Weyant and Hill (1999); Lasky (2003); Stern (2006); Dasgupta (2006); IPCC WG3 (2007); Ackerman (2009).

and to ensure that the economic environment is friendly to innovation. The current international approach in the Kyoto Protocol will be economically costly and have virtually no impact on climate change. In my view, the best approach is also one that is relatively simple—internationally harmonized carbon taxes. Economists and environmentalists will undoubtedly continue to debate the proper level of the carbon price. But all who believe that this is a serious global issue can agree that the current price—zero—is too low and should be promptly corrected.[33]

Rather than delve deeply into this debate here, this section will specifically explore some of the potential costs of delay.

Delay could cause increased costs for a number of reasons. First, if after a period of procrastination it was determined that rapid emissions reductions were in fact required, premature retirement of energy-related capital stock and infrastructure and energy technology retrofits might be necessary. Between today and 2030, energy infrastructure investment decisions alone are expected to total more than $20 trillion dollars.[34] It would be wasteful and costly to scrap these investments. The U.S. National Petroleum Council report underscores the point:

> While uncertainties have always typified the energy business, the risks to supply are accumulating and converging in novel ways. . . . The energy supply system has taken more than a century to build, requiring huge sustained investment in technology, infrastructure, and other elements of the system. Given the global scale of energy supply, its significance, and the time required for substantive changes, inaction is not an option. Isolated actions are not a solution.[35]

Most power plants and other types of energy infrastructure have long lifetimes. Reducing GHG emission levels much below current levels would certainly require a large shift in the pattern of investment. If all new coal-fired power plants were to capture and store CO_2, for example, a large new infrastructure for the transport, distribution, and storage of CO_2 would have to be concurrently developed. Planning in advance for the scale and type of required CO_2 pipeline infrastructure can greatly minimize the combined annualized costs of sequestering a given amount of CO_2 because CO_2 trans-

33. William D. Nordhaus, "'The Question of Global Warming': An Exchange," *New York Review of Books*, September 25, 2008.
34. International Energy Agency (2007).
35. U.S. National Petroleum Council (2007, chapter 2).

port costs can be reduced by building larger-diameter major trunk pipelines that aggregate and transport CO_2 to large storage reservoirs.[36] The IPCC states that the net additional investment required for carbon mitigation ranges from negligible to 5–10 percent, though for some lower-carbon technological options, the costs would certainly be much higher, at least in the short to medium term.[37] In the Stern report, historical precedents for reductions in CO_2 emissions were reviewed, based on work done by the World Resources Institute.[38] During the period 1992–2002, only the economies in transition achieved negative annual growth rates in energy-related CO_2 emissions greater than 1 percent: their negative annual growth rates were minus 3 percent a year. These historical reduction rates stand in stark contrast to the rates identified above by Anderson and Bows as necessary to have a reasonable chance of meeting 450–550 ppm targets.[39]

The problem of premature capital stock retirement is really a problem of "carbon lock-in." Unruh defines carbon lock-in as the "interlocking technological, institutional and social forces that can create policy inertia towards the mitigation of global climate change." This lock-in occurs through a "path-dependent process driven by technological and institutional increasing returns to scale."[40] Energy structures have long lifetimes, ranging from 10–15 years for cars to 20–60 years for energy supply facilities and plants to hundreds of years for buildings. Unless low-carbon technologies are deployed over the next two decades, the world will be locking-in to high enough CO_2 emission levels to break most reasonable emissions budgets. An example of carbon lock-in is related to construction of new fossil fuel power plants between now and 2030. If all the projected new fossil fuel power plants for the world are built without carbon capture and storage, and it turns out that they cannot be easily or cheaply retrofitted to capture CO_2, then the estimated 735 Gt of CO_2 that would be emitted from those new plants alone would account for 47 percent of the total CO_2 budget from 2005 to 2100, for a 450 ppm CO_2 stabilization level.[41] The construction of new buildings represents another form of carbon lock-in, though buildings can be retrofitted with some expenses. It is always cheaper to design new plants and buildings to minimize energy usage and emissions from the start,

36. Middleton and Bielicki (2008).
37. IPCC WG3 (2007, p. 13); Enkvist, Naucler, and Rosander (2007).
38. Stern (2006, p. 179).
39. Anderson and Bows (2008).
40. Unruh (2000, p. 817).
41. Edmonds and others (2007).

though it is not always easy to justify such upfront investments when no climate policies exist.

Procrastination also increases investment uncertainty. Firms and consumers do not know if and when the government will impose laws and regulations related to reducing GHG emissions, and how those regulations will affect them. Financial markets calculate this "regulatory risk," and it is one factor in determining interest rates. As regulatory risks increase, capital costs rise, which slows the pace of energy technology deployment.

Procrastination also could cause increased costs if a suddenly apparent climate crisis dramatically increased demand for low-carbon technologies and services. Abruptly increased demand for low-carbon products and services would drive up prices until suppliers could catch up. If demand persistently escalated faster than supply, then costs could turn out to be substantially higher for carbon mitigation than they might otherwise be through a predictable and steady demand growth that could easily be anticipated and matched by suppliers; this situation, in turn, could cause inflation. Factors that could contribute to higher prices include materials scarcity (for example, silicon for photovoltaic cells), trained-labor scarcity (for example, personnel and firms capable of operating carbon storage facilities), and decommissioning costs.

Related to the previous point, if a rapid pace of emissions reductions were required, the market might be forced to deploy more expensive carbon mitigation technologies. Across the range of potential carbon mitigation technologies, some are assumed to have negative costs (yielding net economic benefits)—usually efficiency technologies—and the supply curve rises indefinitely from there. More economically attractive options may turn out to be too difficult or physically impossible to deploy quickly at very large scale, in which case more expensive options would have to be considered. It should be pointed out that investments in low-carbon energy technology innovation could reduce the risks here because not only should such investments yield new low- or zero-carbon technologies, but they should also help to reduce the costs of existing technologies (see chapter 5).

Finally, there would, of course, be increased costs associated with having to deal with any damages that arise from overshooting the emissions budget and arriving at a much higher GHG concentration level. These can be considered adaptation costs, such as building and repairing levies, moving buildings and other infrastructure back from coastlines, and recovering from more frequent large-scale floods.

The Role for Government

The U.S. government has a clear role to play in the response to climate change because only government can establish the rules of the game for the private sector and society at large. Although the legislative detail can quickly become complex, there are two main steps the government must take to initiate precautionary measures and begin the transition in a timely but deliberate manner. First, the government should set a long-term goal, acknowledging that there will need to be midcourse adjustments given the inherent scientific uncertainties about climate sensitivities, feedbacks, and surprises. By setting the goal, the government implicitly establishes a GHG budget.

Second, the government must place a price on GHG emissions either directly, through a cap-and-trade mechanism or a tax, or, indirectly, through a regulatory regime. A tax, of course, would produce the most predictable and stable price. If a cap-and-trade mechanism is employed, the price will emerge in the emissions-trading market, and it will rise and fall according to demand and supply, thereby being somewhat less predictable. A regulatory regime produces the most uncertain or indirect price because producers and consumers alike must somehow calculate the corresponding carbon abatement costs. The price initially could be relatively modest and then steadily escalate over time. As soon as a price is established for some period lasting into the medium term, a relatively predictable investment climate is created. Project developers can take that carbon price "to the bank" because they can calculate the returns on low-carbon investments with greater precision. The desire for policy certainty is why CEOs testifying before the U.S. Senate Environment and Public Works Committee in June 2007 actually called for mandatory climate policies (U.S. Senate 2007). A price on carbon will have the added benefit of spurring research, development, and demonstration of alternative technologies, and it creates immediate incentives for the deployment of existing low-carbon technologies that were previously uneconomic.

Recent GHG Emission Trends

Even with the entry into force of the Kyoto Protocol in 2005, no country in the world has managed to dramatically reduce its emissions, although some countries have made significant progress. There are many real-world examples of municipalities, individual firms, and state and local governments that have set moderate to ambitious targets and begun to slow, stop, and reverse their GHG

Figure 2-2. Global Carbon Emissions from Fossil Fuel Burning, Cement Manufacture, and Gas Flaring, 1751–2005

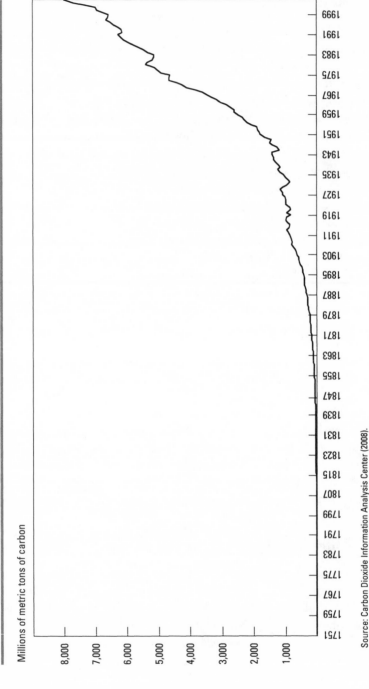

Millions of metric tons of carbon

Source: Carbon Dioxide Information Analysis Center (2008).

emissions—not only in places like Europe and Japan but also within the United States and in some developing countries. Yet the unfortunate reality is that global emissions are still relentlessly rising, as shown in figure 2-2.

Global carbon dioxide emissions from energy and cement production grew 34 percent between 1990 and 2007 according to the Netherlands Environmental Assessment Agency.[42] Between 2000 and 2005, global CO_2 emissions grew 16 percent. Global CO_2 emissions have been growing about four times faster since 2000 than during the previous decade, a matter of some surprise since emissions growth for 2000–07 was above even the most extreme fossil fuel–intensive emissions scenario of the IPCC Special Report on Emissions Scenarios.[43]

The UN Framework Convention on Climate Change (UNFCCC) was negotiated in 1992 at the Rio Earth Summit. This convention, signed and ratified by 192 countries, including the United States, created an overall global framework for cooperation among nations to tackle the threat of climate change. The UNFCCC contains a number of negotiated principles, and it committed industrialized countries to adopt national policies to limit anthropogenic emissions of GHGs and to "aim" to reduce GHG emissions to 1990 levels. All countries, industrialized and developing alike, were expected to develop and produce an inventory of GHG emissions, report on national policies, and cooperate in preparing for adaptation to global climate change. One important principle established was that there were "common but differentiated" responsibilities among nations. The industrialized countries had accepted the obligation to take the lead in reducing emissions.

Despite the UNFCCC's stated aim of reducing GHG emissions to 1990 levels, no industrialized country made serious efforts to reduce emissions after the framework convention entered into force. But soon after the Intergovernmental Panel on Climate Change released its 1995 assessment report, there were calls for a more stringent, legally binding international agreement, and negotiations commenced for what became the Kyoto Protocol to the UNFCCC.

The Kyoto Protocol negotiations concluded in December 1997 in Japan. Many countries, including the United States, quickly signed the Protocol, but not all the signatories went on to ratify the Protocol. Conspicuously, of the

42. Netherlands Environmental Assessment Agency (NEAA), "Global CO_2 Emissions from Fossil Fuels and Cement Production by Region, 1990–2007" (www.mnp.nl/en/publications/2008/Global CO2emissionsthrough2007.html [September 2008]).

43. Global Carbon Project, "Growth in the Global Carbon Budget" (www.globalcarbonproject. org/global/pdf/Press%20Release_GCP.pdf [September 2008]). See also IPCC (2000).

183 countries who signed the Kyoto Protocol, the United States and Kaza-khstan are the only two who still have not ratified it. In other words, the United States is the only major industrialized country that has not ratified the Protocol, even though it actively participated in its negotiations and ratified the UNFCCC.

The Kyoto Protocol set forth a number of commitments for industrialized countries, most notably that thirty-seven industrialized countries and the European Community are subject to binding targets for reducing GHG emissions 5 percent below 1990 levels between 2008 and 2012. Developing countries are not subject to binding emissions reduction targets during this period. The Protocol also created a number of new mechanisms, including a system for international emissions trading and the Clean Development Mechanism, whereby an industrialized country can pay for emissions reductions in a developing country and take credit for those reductions at home.

The good news is that of all the industrialized countries that ratified the Kyoto Protocol, twenty-two had reduced their GHG emissions somewhat below 1990 levels by 2005, according to the latest data available.[44] Three countries had halted the growth in their emissions at 1990 levels, and the emissions of sixteen industrialized countries were still growing. The worst offender, Turkey, experienced a 75 percent increase in GHG emissions between 1990 and 2005. Australia, New Zealand, and Canada all increased their emissions 25 percent during the same time period, while Japan increased 7 percent, and U.S. emissions grew 16 percent. Many of the countries that reduced their emissions had special circumstances, especially the economies in transition, including the Russian Federation, Czech Republic, Poland, and the Ukraine. These countries experienced economic collapse, so their emissions had declined far below the target they assumed under the Kyoto Protocol. The United Kingdom reduced its emissions 15 percent, and Germany 18 percent, partly due to economic restructuring in those countries and partly due to significant efforts to improve energy efficiency and expand the use of renewable energy. In addition, there are some countries that achieved significant reductions without extenuating circumstances, notably Denmark and Sweden.

The United Kingdom adopted a Kyoto target of 12.5 percent reduction below 1990 levels by 2012. The House of Lords approved a climate change bill in March 2008 that contained further targets of 26 percent and 60 percent below 1990 GHG emission levels by the years 2020 and 2050, respectively. This bill is currently under consideration by the House of Commons. The United

44. Excluding emissions from land-use change and forestry (UNFCCC 2007).

Kingdom experienced fairly substantial GHG emissions reductions of approximately 1.5 percent a year between 1990 and 2000, but since then the United Kingdom has only managed to hold emissions more or less constant.[45] Altogether, the United Kingdom achieved emissions reductions of 15.7 percent below 1990 levels by 2005. Major factors in the UK success were the liberalization of the energy market, which in turn caused shifts to cleaner fuels for power production (mainly from coal to gas but also major expansion of wind power), and significant reductions in HCFC production and methane from landfills. There are currently 169 onshore and 7 offshore wind farms in the United Kingdom, with another 33 onshore and 7 offshore wind farms under construction.[46] The United Kingdom is on target to be well below its Kyoto commitment by 2010.[47]

Denmark assumed a very ambitious target of reducing its GHG emissions 21 percent below 1990 by 2012 under the Kyoto Protocol. As of 2005, Danish emissions were 7 percent below 1990 levels due to emissions reductions from the power sector, agricultural soils, and households. Denmark has become famous for its aggressive shift into wind power, making use of offshore wind resources. As of 2006, Denmark obtained 26 percent of its electricity from renewable sources.[48] Offsetting the impressive gains in the Danish electricity sector is the growth in emissions from the transportation sector there, a common challenge in many countries, including the United States. With all the steps that Denmark has taken, it does not appear that it will meet its Kyoto target with domestic measures alone. Denmark currently plans to purchase some emissions credits from other countries through the European Emissions Trading Scheme and the Clean Development Mechanism.[49]

Numerous multinational firms have adopted voluntary emissions reduction targets during the past decade, and some, of course, have been forced to reduce emissions if they operate in countries that have ratified the Kyoto Protocol and have passed implementing legislation. Exelon, one of the largest U.S. electricity utilities in the United States, pledged in February 2008 to cut or offset all of its emissions by 2020, approximately 15 million metric tons of CO_2e a year, at a cost of $10 billion in 2012 dollars.[50] Xerox achieved an 18 percent

45. U.K. Department for Environment, Food and Rural Affairs (2006, section 2, p. 27).

46. British Wind Energy Association, "Statistics" (www.bwea.com/statistics [September 5, 2008]).

47. European Environment Agency, "Greenhouse Gas Emission Trends and Projections in Europe 2007," Report 5-2007 (http://reports.eea.europa.eu/eea_report_2007_5/en).

48. Renewable Energy Policy Network for the Twenty-First Century (2008).

49. European Environment Agency, "Greenhouse Gas Emission Trends."

50. Exelon, "Exelon 2020: A Low Carbon Roadmap, Fact Sheet" (www.exeloncorp.com/NR/rdonlyres/9727BE9B-9BE4-44F2-8268-1132F5F27D14/0/Exelon2020FactSheet.pdf. [September 2008]).

reduction in its GHG emissions six years early. General Motors pledged to reduce its North American GHG emissions by 40 percent from 2000 to 2010 and already achieved an initial goal of reducing North American GHG emissions by 23 percent by 2005.[51]

Even though the Kyoto Protocol catalyzed significant efforts to reduce emissions by many industrialized countries (that have ratified the agreement), the agreement has been criticized by some, especially in the United States, as being too ineffective or weak as compared with the magnitude of the need to reduce emissions. The argument that it is too weak stems from the fact that industrialized countries are only required to reduce emissions slightly below 1990 levels whereas, for the time being, developing country emissions are allowed to continue to rise. Others have pointed out that since neither of the two largest aggregate emitters, the United States and China, are committed to binding targets under the Kyoto Protocol (the United States because it failed to ratify, China because it is a developing country), nearly half of global emissions are growing unabated. China's aggregate emissions surpassed U.S. total emissions in 2006, and as of 2007, the two countries together accounted for 46 percent of global CO_2 emissions.

The United States and China

The United States and China are the two countries with the unique ability to make or break the climate change threat. Given how large both are in terms of their aggregate emissions, if either one fails to effectively manage its GHG emissions during this century, it really will be impossible to substantially reduce the threat of climate change. If both fail, the game is over. This section explores what the two largest emitters in the world are doing to manage or reduce their GHG emissions.

The two countries share many of the same energy-climate challenges, especially their heavy dependence on coal, the most carbon-intensive fossil fuel, and the large and growing emissions from their transportation sectors. But, of course, the two countries are very different as well. It is worth exploring how these two countries compare in energy terms. While China is quickly closing the gap with regard to total energy consumption, its energy use is three-quarters the level of U.S. usage. China's oil imports have grown more rapidly than U.S. imports, but, overall, they are just one-third the total level of U.S. oil

---

51. U.S. Environmental Protection Agency, "Climate Leaders: Partners" (www.epa.gov/stateply/partners/index.html).

imports. China's electricity capacity is still smaller than U.S. capacity but growing at an astonishing rate, whereas the U.S. electricity system is growing slowly, in more of a replacement mode. China's coal consumption is twice as large as U.S. coal consumption, and this is largely because coal is the resource in greatest abundance in China, even though the United States has much larger coal reserves. In terms of passenger cars, the United States has 230 million cars, light trucks, and SUVs, whereas China has approximately 38 million. Total carbon dioxide emissions are about equal, though China's GHG emissions surpassed U.S. emissions in 2007.[52] On a per capita basis, however, China's emissions are just one-fifth the size of U.S. emissions.

Despite the fact that the United States has not ratified the Kyoto Protocol, the federal government has enacted numerous policies that have had the indirect effect of reducing the growth of GHG emissions in the United States. Growth in U.S. CO_2 emissions has slowed considerably since 2000, essentially leveling off at about 6 billion tons per year during this century.[53] The strengthening of the Corporate Average Fuel Economy standards in the Energy Independence and Security Act of 2007 will minimize growth or possibly reduce U.S. transportation-related CO_2 emissions during the next decade.

Meanwhile, Congress has earnestly begun consideration of several climate change bills. In December 2007, the Senate Environment and Public Works Committee voted along party lines to approve a bill that would have capped U.S. GHG emissions in 2012 and reduced them 60 percent by 2050. In June 2008, the Senate formally took up the Lieberman-Warner Climate Security Act, the first time a climate change bill had come to the floor of the Senate. While the Senate majority leader decided to end debate on the bill after just one week, the opportunity to begin debate was a major step forward. There was less legislative activity in the House of Representatives during 2008.

Not only have policies been enacted at the federal level, but states and municipalities across the United States have also passed legislation to promote renewable energy, improve energy efficiency, and in some cases to explicitly reduce GHG emissions through regulation and GHG cap-and-trade programs. Ten northeastern and mid-Atlantic states created the Regional Greenhouse Gas Initiative, the first U.S. cap-and-trade system aimed at reducing CO_2 emissions from power plants. The program begins by capping emissions in 2009, and then reducing emissions 10 percent by 2019.

52. NEAA, "Global CO_2 Emissions."
53. U.S. Environmental Protection Agency (2008).

In 2007 six U.S. states and one Canadian province established the Midwestern Regional Greenhouse Gas Reduction Accord, which established a long-term target of emissions 60 to 80 percent below 2007 levels, and set in motion the creation of a multisector cap-and-trade system for the region. Also in 2007, seven western U.S. states and two Canadian provinces created the Western Climate Initiative, which has a regional target of a 15 percent reduction below 2005 levels by 2020. Seventeen states have set GHG reduction targets through legislation or executive order or both. In September 2006, California governor Arnold Schwarzenegger signed AB 32, the Global Warming Solutions Act, which caps California's GHG emissions at 1990 levels by 2020. This legislation is the first enforceable statewide program in the United States that includes penalties for noncompliance.[54]

China has a very large population of 1.3 billion people as of 2007, accounting for 20 percent of the world's total. Per capita income in 2007 was $2,461, with approximately 30 million people living in poverty.[55] Total energy consumption in China increased 70 percent between 2000 and 2005, and total coal consumption increased by 75 percent during the same time period.[56] This astonishing rate of growth indicates that China's entire energy system is doubling in size every five years so far this century. But even China, with no formal obligations to reduce GHG emissions under the Kyoto Protocol, has taken many steps to reduce emissions through policies and measures to boost energy efficiency and renewable energy.

Because so much of China's energy depends on coal, efficiency measures that result in reduced coal combustion will greatly help to reduce GHG emissions. Between 1980 and 2004, China's energy intensity (the amount of energy used to generate economic activity—usually expressed as total energy consumption divided by GDP) declined dramatically. The Chinese government's eleventh five-year plan (2006–10) calls for a further 20 percent reduction in energy intensity from 2005 levels by 2010. But the goal is proving to be challenging: according to Chinese government statistics, energy intensity dropped 1.8 percent in 2006 and 3.7 percent in 2007, well short of the 5 percent a year target.[57]

The energy intensity target was also at the heart of the voluntary climate change plan that the Chinese government announced in June 2007 and revised in October 2008.[58] By improving thermal efficiency in power plants,

54. Pew Center on Global Climate Change (2007).
55. People's Republic of China (PRC, 2008).
56. World Bank and China State Environmental Protection Administration (2007).
57. PRC (2008).
58. Ibid.

for example, the government estimated that it could reduce carbon dioxide emissions by 110 million tons by 2010 cumulatively.[59] Notably, the Chinese government also issued its first fuel efficiency standards for passenger cars in 2005, and these were strengthened in 2008. China also implemented vehicle excise taxes so that if a buyer purchases a car or SUV with a big engine, that individual will pay a much higher tax than if she or he purchased a car with a small, energy-efficient engine. These automotive efficiency policies are considerably more stringent than comparable ones in the United States. The Chinese government also adopted aggressive efficiency standards for appliances. The China Energy Group (2001) at Lawrence Berkeley National Laboratory estimates that those standards are already reducing GHG emissions in China substantially and that even more stringent minimum appliance standards could reduce carbon emissions by 19 million metric tons a year by 2020.[60]

The Chinese government has also aggressively promoted low-carbon energy supply options, especially renewable energy, hydropower, and nuclear energy. China has twice as much installed renewable power capacity as the United States.[61] In fact, China led the world in terms of total installed capacity of renewable energy at 42 gigawatts (GW; compared with 23 GW in the United States) as of 2005. That year, China accounted for 63 percent of the solar hot water capacity in the world and had installed 1.3 GW of wind capacity. China also passed the Renewable Energy Law in 2005, which requires grid operators to purchase electricity from renewable generators, and China set a target of having 10 percent of its electric power generation capacity come from renewable energy sources by 2010 (not including large hydropower). By expanding renewable energy (including bioenergy), the Chinese government estimated it could reduce carbon dioxide emissions by 90 million tons by 2010. China has exploited its large hydropower resources at some social and ecological cost due to forced relocations of communities, loss of ecosystems, and decreased river flow, but it believes it has substantial scope for increasing small-scale hydropower, and in fact, it estimated that it could achieve a reduction of 500 million tons of carbon dioxide by 2010 via increased use of small-scale hydropower. Compared to its use of coal and hydropower, however, China has scarcely begun its expansion of nuclear power. By 2020 the Chinese government plans to have built 40–60 GW of new nuclear power capacity, but

59. National Development and Reform Commission, People's Republic of China, "China's National Climate Change Programme," June 2007 (www.ccchina.gov.cn/WebSite/CCChina/UpFile/File188.pdf).
60. Lin (2006).
61. Excluding large hydropower but including small hydropower.

even if it succeeds in expanding so rapidly, the 40 GW would only account for about 4 percent of the anticipated total capacity by then.

Because of China's heavy reliance on coal and its current stage of industrialization and economic development, China's emissions have grown rapidly so far this century. Between 2000 and 2007, China's CO_2 emissions grew 14 percent a year, on average, according to the Netherlands Environmental Assessment Agency.[62] It will be extremely difficult for China to reduce emissions in absolute terms any time soon because of its current stage of industrialization and heavy reliance on coal. Coal still accounts for 80 percent of China's energy supply and GHG emissions.

Each new coal-fired power plant that is built represents a fifty- to seventy-five-year commitment because these plants are unlikely to be prematurely retired. The International Energy Agency estimates that 55 percent of the new coal-fired power plants that will be built between now and 2030 will be built in China.[63] By using the cheapest technologies currently available for coal-fired power plants and industrial facilities (which is perfectly rational in strict economic terms), China and the United States are effectively locking themselves and the world into high GHG emissions. The rapid growth in power plants and related infrastructure in China is expected to continue, so "leapfrogging" to lower-carbon technologies in the near term is absolutely critical. But technological leapfrogging is not an automatic process because developing countries either lack the technological capabilities or cannot afford the costs of more advanced technologies.[64] Will all these new plants utilize conventional high-carbon technologies or best available low-carbon technologies? At China's recent growth rate, it will have installed the same amount of electricity capacity as the United States currently has (1,089 GW) within five to ten years, virtually all of it in conventional coal-fired power. Unfortunately, the incremental costs of advanced coal technologies in the domestic Chinese context appear to be substantial, not including the costs of carbon capture and storage.[65] Thus research, development, and demonstration in carbon capture and storage is urgently needed (see chapter 3).

Since the United States also is not building advanced coal-fired power plants that capture and store CO_2, it is especially hard to make the case that China should start building such plants now despite the environmental imperative to do so. Thus the United States not only needs to start slowing,

62. NEAA, "Global CO_2 Emissions."
63. International Energy Agency (2007).
64. Gallagher (2006).
65. Zhao and others (2008).

stopping, and reversing its own emissions, but it will almost certainly have to help China to reduce its emissions as well if the climate change threat is to be effectively tackled.

Conclusion

One cannot help but notice the widening discrepancy between the apparent scale of the challenge of avoiding dangerous climate change and current policies already enacted or being seriously considered, especially in the United States and China. Despite the many efforts being made in both countries and internationally to improve energy efficiency, exploit renewable energy, and invest in energy technology innovation, global GHG emissions are still growing rapidly. The potential consequences of procrastination are deeply worrisome, but there is little evidence that the largest emitters have yet considered how short the window of opportunity may be to act in time.

The United States needs to set a long-term target and thereby establish a national emissions budget. The government must enact a set of concrete, mandatory, enforceable carbon mitigation policies domestically that impose a clear cost on carbon dioxide emissions. And the Obama administration urgently needs to begin direct talks with the Chinese government about how to collaborate to reduce the threat of global climate disruption. As a much richer country (and much larger emitter in terms of its cumulative historical emissions), the United States has an obligation to take the lead. Leading by example might be the best way to inspire the Chinese to take responsible steps to do their fair share as well.

References

Ackerman, Frank. 2009. *Can We Afford the Future? The Economics of a Warming World.* London: Zed Books.

Anderson, Kevin, and Alice Bows. 2008. "Reframing the Climate Change Challenge in Light of Post-2000 Emission Trends." *Philosophical Transactions of the Royal Society A: Mathematical, Physical and Engineering Sciences* 366, no. 1882: 3863–82.

Bazerman, Max H., and Michael D. Watkins. 2004. *Predictable Surprises: The Disasters You Should Have Seen Coming and How to Prevent Them.* Harvard Business School Press.

Bows, Alice, and Kevin Anderson. 2007. "A Bill We Can't Pay." *Parliamentary Brief* (May).

Bows, Alice, and others. 2006. "Living within a Carbon Budget." University of Manchester, Tyndall Center for Climate Change Research (July).

Cagin, Seth, and Phillip Dray. 1993. *Between Earth and Sky: How CFCs Changed Our World and Endangered the Ozone Layer*. New York: Pantheon.

Carbon Dioxide Information Analysis Center. 2008. "Global CO_2 Emissions from Fossil-Fuel Burning, Cement Manufacture, and Gas-Flaring, 1751–2005."Oak Ridge, Tenn.: Oak Ridge National Laboratory (August).

China Energy Group. 2001. *China Energy Data Book*. Version 6.0. Berkeley, Calif.: Lawrence Berkeley National Laboratory. CD.

Chylek, Petr, Manvendra K. Dubey, and Glen Lesins. 2006. "Greenland Warming of 1920–1930 and 1995–2005." *Geophysical Research Letters* 33: L11707.

Committee on Abrupt Climate Change, National Research Council. 2002. *Abrupt Climate Change: Inevitable Surprises*. Washington: National Academy Press.

Dasgupta, Partha. 2006. "Comments on the Stern Review's Economics of Climate Change." Prepared for seminar on the Stern Review's Economics of Climate Change, Foundation for Science and Technology, Royal Society, London, November 8.

Edmonds, Jae A., and others. 2007. "Global Energy Technology Strategy: Addressing Climate Change. Phase 2 Findings. Executive Summary." Richland, Wash: Pacific Northwest National Laboratory.

Enkvist, Per-Anders, Tomas Naucler, and Jerker Rosander. 2007. "A Cost Curve for Greenhouse Gas Reduction." *McKinsey Quarterly*, no. 1.

Gallagher, Kelly Sims. 2006. "Limits to Leapfrogging in Energy Technologies: Evidence from the Chinese Automobile Industry." *Energy Policy* 34, no. 4: 383–94.

Hansen, James, and others. 2008. "Target Atmospheric CO_2: Where Should Humanity Aim?" *Open Atmospheric Science Journal* 2: 217–31 (www.bentham.org/open/toascj/openaccess2.htm).

Hassan, Rashid, Robert Scholes, and Neville Ash, eds. 2005. *Ecosystems and Human Well Being*, vol. 1: *Current State and Trends*. Washington: Island Press.

Holdren, John P. 2008. "Science and Technology for Sustainable Well-Being." *Science* 319, no. 5862: 424–34.

International Energy Agency. 2007. *World Energy Outlook 2007: China and India Insights*. Paris.

Intergovernmental Panel on Climate Change (IPCC). 2000. *Emissions Scenarios. A Special Report of Working Group III of the Intergovernmental Panel on Climate Change*. Geneva.

———. 2007. *Climate Change 2007: Synthesis Report. Contribution of Working Groups I, II and III to the Fourth Assessment Report of the Intergovernmental Panel on Climate Change*. Geneva.

———. Working Group I (IPCC WG1). 2007. *Climate Change 2007: The Physical Science Basis. Contribution of Working Group I to the Fourth Assessment Report of the Intergovernmental Panel on Climate Change*. Cambridge University Press.

———. Working Group III (IPCC WG3). 2007. "Summary for Policymakers." In *Climate Change 2007: Mitigation. Contribution of Working Group III to the Fourth*

Assessment Report of the Intergovernmental Panel on Climate Change. Cambridge University Press.

Lasky, Mark. 2003. "The Economic Costs of Reducing Emissions of Greenhouse Gases: A Survey of Economic Models." Technical Paper 2003-03. Congressional Budget Office (May).

Lin, Jiang. 2006. "Mitigating Carbon Emissions: The Potential of Improving Energy Efficiency of Household Appliances in China." Report 60973. Berkeley, Calif.: Lawrence Berkeley National Laboratory.

Meinshausen, Malte. 2006. "What Does a 2°C Target Mean for Greenhouse Gas Concentrations? A Brief Analysis Based on Multi-Gas Emission Pathways and Several Climate Sensitivity Uncertainty Estimates." In *Avoiding Dangerous Climate Change*, edited by Hans J. Schellnhuber and others, pp. 253–79. Cambridge University Press.

Middleton, Richard S., and Jeffrey M. Bielicki. 2009. "A Scalable Infrastructure Model for Carbon Capture and Storage: SimCCS." *Energy Policy* 37: 1052–60.

Pearce, David W., and others. 1996. "The Social Costs of Climate Change: Greenhouse Damage and the Benefits of Control." In *Climate Change 1995: Economic and Social Dimensions of Climate Change*, edited by James P. Bruce, Hoesung Lee, and Erik F. Haites, pp. 179–224. Cambridge University Press.

People's Republic of China (PRC). State Council. 2008. "White Paper: China's Policies and Actions on Climate Change." Beijing: Foreign Languages Press (October).

Pew Center on Global Climate Change. 2007. "Learning from State Action on Climate Change." Update. Arlington, Va. (March).

Renewable Energy Policy Network for the Twenty-First Century (REN21). 2008. *Renewables 2007 Global Status Report.* Paris: REN21 Secretariat, and Washington: Worldwatch Institute.

Stavins, Robert. 1999. "The Costs of Carbon Sequestration: A Revealed-Preference Approach." *American Economic Review* 89, no. 4: 994–1009.

Stern, Nicholas. 2006. *Stern Review on the Economics of Climate Change.* Cambridge, UK: HM Treasury.

U.K. Department for Environment, Food and Rural Affairs. 2006. *Climate Change: The UK Programme 2006.* Norwich, U.K.: Office of Public Sector Information.

UN Framework Convention on Climate Change (UNFCCC). 2007. "National Greenhouse Gas Inventory Data for the Period 1990–2005." Report FCCC/SBI/2007/30 (October).

Unruh, Gregory C. 2000, "Understanding Carbon Lock-In." *Energy Policy* 28, no. 12: 817–30.

U.S. Environmental Protection Agency. 2008. "Inventory of U.S. Greenhouse Gas Emissions and Sinks: 1990–2006." USEPA 430-R-08-005 (April).

U.S. National Petroleum Council. 2007. *Hard Truths: Facing Hard Truths about Energy.* Washington.

U.S. Senate. Committee on Environment and Public Works. 2007. "Testimony of Peter A. Darbee, Chairman, CEO and President, PG&E Corporation." *Hearing on Examining Global Warming in the Power Plant Sector.* 110 Cong. 1 sess. (June 28).

Wang, Tao, and James Watson. 2008. "Carbon Emissions Scenarios for China to 2100." Tyndall Working Paper 121. University of Manchester, Tyndall Center for Climate Change Research.

Westerling, Anthony L., and others. 2006. "Warming and Earlier Spring Increase Western U.S. Forest Wildfire Activity." *Science* 313, no. 5789: 940–43.

Weyant, John P., and Jennifer N. Hill. 1999. "Introduction and Overview." *Energy Journal* (Special Issue: *The Costs of the Kyoto Protocol: A Multi-Model Evaluation*), pp. vii–xliv.

World Bank and China State Environmental Protection Administration. 2007. "Cost of Pollution in China: Economic Estimates of Physical Damages." Washington: World Bank (February).

Zhao, Lifeng, and others. 2008. "Economic Assessment of Deploying Advanced Coal Power Technologies in the Chinese Context." *Energy Policy* 36, no. 7: 2709–18.

three
Making Carbon Capture and Storage Work

Daniel P. Schrag

President Obama faces an old challenge of creating a national energy policy. That policy will be designed with multiple objectives in mind. After a year of record oil prices that added to U.S. economic troubles, some want an energy policy that will maintain lower energy prices. With nearly 150,000 troops still in Iraq and tensions raised with Russia over the Georgian invasion, some want an energy policy that will reduce American dependence on fossil fuel imports from these and other geopolitically sensitive regions. And with atmospheric carbon dioxide now more than 385 parts per million and rising, some want an energy policy that will reduce greenhouse gas emissions.

This chapter focuses on how the United States can accomplish the third objective, reducing carbon dioxide emissions from fossil fuels. I argue that demonstration and deployment of technologies to capture carbon dioxide from large stationary sources, storing the waste CO_2 in geological formations, is likely to be an essential component of any carbon reduction strategy, both for the United States and for the world, and is also consistent with economic and security concerns. It also reviews the major technical challenges involved with widespread deployment of carbon capture and storage, and discusses policies that would lead to the specific goal of capturing and storing the CO_2 from all large stationary sources by the middle of this century.

Several excellent reviews of carbon capture and storage have appeared in recent years, in particular the MIT report on *The Future of Coal* and the *IPCC*

Special Report on Carbon Dioxide Capture and Storage.[1] Topics in this chapter are discussed in much greater detail in these reports and the references therein.

The Case for Carbon Capture and Storage

Strategies to lower CO_2 emissions to mitigate climate change can be grouped into three broad categories. The first category involves reducing CO_2 emissions by reducing energy consumption. This does not necessarily mean reducing energy use by reducing economic activity through conservation but also by investing in low-energy social adaptations, such as public transportation systems, or by adopting energy-efficient technologies in buildings, in automobiles, and throughout the economy. Huge discrepancies in energy efficiency exist today between countries, even within the developed world; in general, countries that have higher historical energy prices, including many in western Europe, are more efficient than those countries with inexpensive energy, although the differences can also be explained by historical investments in cities and suburbs, in highways and public transportation systems, government policies, and a variety of other factors. But whatever the cause of the current differences between countries, there is great potential across the developed and the developing world to dramatically lower energy use through smarter and better energy systems.[2]

The second category involves expansion of nonfossil energy systems including wind, solar, biomass, geothermal, and nuclear power. Wind is currently the most economical of these systems for electricity generation, at least in appropriate areas.[3] However, wind requires excess capacity because of intermittency in the wind resource, and so it is difficult to deploy as a source for base load power unless storage technologies become better and cheaper. Solar generated electricity has similar energy storage issues and is also expensive compared with wind or nuclear power, although solar thermal plants may be an interesting alternative to photovoltaic devices in addressing these concerns. Nuclear power can be used for base load power, unlike wind or solar photovoltaic power, but it has issues of safety and storage and handling of nuclear waste, and there are security concerns regarding nuclear weapons proliferation that must be addressed before widespread expansion is likely, at

1. See Deutch and Moniz (2007) and Intergovernmental Panel on Climate Change (IPCC 2005), respectively.
2. United Nations Foundation (2007).
3. See, for example, Bird and Kaiser (2007).

least in the United States and western Europe outside of France. Because of these factors, as well as other regulatory uncertainties, the economic cost of new nuclear power plants in the United States remains uncertain.[4]

Outside of the electric realm, biomass converted to transportation fuel may play a major role in reducing CO_2 emissions in the transportation sector, at least until powerful, inexpensive, and reliable battery technologies or some alternative transportation technologies are developed. For example, Brazil currently obtains most of its transportation fuel from sugarcane fermentation into ethanol, and programs around the world are following suit.[5] A more efficient technology may be the conversion of biomass into synthetic diesel and jet fuel via the Fischer-Tropsch process used by the Germans in World War II to make coal into liquid fuel. This process has the advantages of creating a more diverse range of fuel products, including jet fuel for air transport, and of being more efficient through use of all types of biomass, not just sugar (or cellulose for a cellulosic conversion process). Moreover, the Fischer-Tropsch process, which involves gasification of the biomass followed by conversion to liquid fuel via a cobalt or iron catalyst, requires removal of CO_2 to avoid poisoning the catalyst, making it easily adapted to capture and storage of CO_2, as discussed below.

The third category involves CO_2 capture from emissions sources and then storage in geological repositories, often referred to as carbon capture and storage (CCS). CCS appears particularly attractive because it has the potential to allow some of the largest economies of the world to use their abundant and inexpensive coal resources without releasing vast amounts of CO_2 to the atmosphere. Coal produces the most CO_2 per unit energy of all fossil fuels, nearly twice as much as natural gas. And unlike petroleum and natural gas, which are predicted to decline in total production well before the middle of the century, there is enough coal to last for centuries, at least at current rates of use, and that makes it inexpensive relative to almost every other source of energy. Even with huge improvements in efficiency and increases in nuclear, solar, wind, and biomass power, the world is likely to depend heavily on coal, especially the five countries that hold 75 percent of world coal reserves: United States, Russia, China, India, and Australia. Domestic use of coal in the United States is also advantageous from a national security perspective, as the main alternative to coal for power generation is natural gas, which, in the future, may mean greater reliance on imports of liquefied natural gas from the Middle East and Russia.

---

4. Deutch and Moniz (2003).
5. Gallagher and others (2006).

The vast majority of coal use is for electricity generation. In the United States, for example, 95 percent of coal in 2007 was used for power plants or for combined heat and power. Although coal-fired power plants make up only a third of the U.S. generating capacity, they account for 50 percent of total electricity generation and 40 percent of CO_2 emissions. This is due to the low price of electricity generated from coal plants and the operational difficulties of turning the plants on and off, making them the backbone of the U.S. power generating system. Incorporating CCS for these power plants could reduce their CO_2 emissions by at least 80 percent, enabling the United States to continue to use its vast coal reserves without harming the climate system.

Another reason why some have found CCS to be so compelling as part of a climate change strategy is that CCS, if cost effective, might allow the world to transition to a low-carbon economy without discarding capital investments that have been made in electricity infrastructure. In 2007 there were 2,211 power plants that emitted at least 1 million tons of CO_2 a year: 1,068 were in Asia (559 in China), 567 in North America (520 in the United States), 375 in Europe, and 157 in Africa.[6] Together these power plants released 10 billion tons of CO_2, or one-third of global emissions. To the extent that some of these plants can be retrofitted with capture technology and that appropriate storage locations can be identified, CCS would allow the world to continue to use these facilities for many decades but dramatically reduce their environmental impact.

Another consideration is the timescale over which it is possible to build new energy systems. Eliminating carbon emissions from electricity generation with new nuclear power plants, for example, would require building two large plants each week for the next 100 years. This rate of change seems improbable given current constraints on steel production, construction capacity, and education of operators, as well as many other practical considerations. Given the capital investment the world has already made in power generation, CCS appears to be one of the few ways to lower carbon emissions and still make use of those investments.

Although coal is the major motivation for the development of CCS, it is important to note that CCS need not apply exclusively to coal; any point source of CO_2 can be sequestered, including CO_2 from combusted biomass, which would result in negative emissions. Indeed, the oldest CCS installation, StatoilHydro's Sleipner project that started in 1996, sequesters CO_2 sep-

6. Carbon Monitoring for Action, "Power Plant Data" (http://carma.org/dig).

arated from natural gas and injects it into a sandstone formation off the Norwegian coast. Even if one could imagine a U.S. energy system that did not rely heavily on coal for electricity (which is difficult), CCS would still be needed to achieve carbon reduction goals for large natural gas power plants, oil refineries, smelting and steel manufacturing facilities, fertilizer plants, and ethanol distilleries. Indeed, some of these noncoal stationary sources of CO_2 emissions are likely to be the first targets for CCS deployment as some of them already produce concentrated CO_2, eliminating most of the capture costs. Of course, limiting the amount of future coal use would greatly diminish the required scale of CCS, but CCS is likely to represent a large portion of any global strategy to lower CO_2 emissions, at least for the next century, even without new investments in coal-fired generating plants.

A final reason why CCS is so important for U.S. energy policy to achieve a dramatic reduction in global CO_2 emissions is that CCS is likely to be extremely important in China, India, and Russia as they all have large coal reserves. Widespread adoption of CCS in the United States over the next few decades will make it more likely that similar systems will be deployed overseas, especially in the rapidly growing economies with high present and future CO_2 emissions.

Scientific and Technological Challenges

The scientific and technological challenges associated with carbon capture are quite distinct from the challenges related to carbon storage. In the following sections, some of the most serious obstacles to CCS deployment for capture and for storage are discussed.

Capture

Geological storage of CO_2 requires a relatively pure gas stream; the energetic costs of compression prohibit sequestration of dilute mixtures. Some industrial processes release concentrated CO_2 and are therefore perfectly suited for storage. These include ammonia plants (Haber-Bosch process) and synthetic fuel plants (Fischer-Tropsch process) that capture CO_2 because it interferes with the main catalytic reactions. Some components of petroleum refineries, in particular CO_2 released from the production of hydrogen through steam methane reforming, are capable of producing pure CO_2 streams in a similar fashion. A range of capture technologies exists for such plants, most of which are based on chemical solutions (for example, Selexol) that absorb CO_2 at high pressure inside the reactor and release it at atmospheric pressure. This means that the only added cost for these facilities to make them ready for CO_2 storage, aside from the actual injection and storage costs, is compression for transport via pipelines. At smaller scales, many ethanol plants release concentrated CO_2 from the fermentation process that is similarly at low pressure and must be compressed before being transported to a storage site. All of these facilities are excellent prospects for early deployment of CCS because the compression, transportation, and injection costs are relatively minor and will likely be economical under a cap-and-trade system with only a modest price on CO_2 (less than $20 per ton).[7]

For power plants, which make up the largest class of CCS targets, the capture of CO_2 is very expensive, both in capital expenditures and in energy costs, in part because CO_2 capture is an additional cost and not built into the overall process. Conventional pulverized coal (PC) plants, whether critical, supercritical, or ultra-supercritical, burn coal in air, producing a low-pressure effluent composed of roughly 10 percent CO_2 in nitrogen. CO_2 can be scrubbed from the nitrogen using amine liquids, chilled ammonia, or other materials with high affinity for CO_2. These various CO_2 "scrubbers" must then be regenerated by heating them to release the CO_2, which can use a significant amount of energy. Overall, the energy penalty associated with capture and storage represents roughly 30 percent of the electricity from an average plant in the United States and may raise the generating cost of electricity from coal by 50 percent or more.[8] New plants designed with CCS in mind will likely be much more economical than retrofitting existing plants, as discussed below.[9]

If a PC plant, whether new or old, is outfitted with postcombustion capture technology, the energy to operate the scrubbing systems (primarily as heat to remove CO_2 from the amine or chilled ammonia scrubbers) as well as the electricity to run the gas compressors will reduce the total electrical output of the plant. Better scrubbing materials are being developed that are able to release CO_2 at lower temperatures, allowing the possibility that much of the energy used in the capture process can be waste heat from the coal combustion, minimizing the drop in electrical output. Such improvements in the ability to employ waste heat in the capture process will be critical in driving down the overall cost of CCS.

When one considers adding CCS to existing PC plants in the United States, the energy penalty entailed will constrain the extent to which CCS will be deployed as a retrofit. The average efficiency of coal-fired power plants in

---

7. McCoy and Rubin (2005).
8. Anderson and Newell (2004); IPCC (2005).
9. See also Rubin, Chen, and Rao (2007); Bohm and others (2007).

the United States is 32 percent, but with a wide range between 20 and 40 percent. By definition, a plant with low efficiency will use more coal than a high-efficiency plant to produce the same amount of power and therefore will produce more CO_2. This means that a PC plant with low efficiency will have to spend a much higher fraction of the electricity it produces on capture and compression than a higher-efficiency plant.[10] Indeed, for the least efficient quartile of PC plants in the United States, retrofitting with CCS equipment is unlikely to make economic sense under any foreseeable cap-and-trade regime since the energy penalty is simply too high. How to decommission these older, low-efficiency power plants will someday be a major challenge for U.S. energy policy, as discussed below.

Another way to capture CO_2 from a power plant is through what is commonly called precombustion technologies. One type of precombustion system uses an oxyfuel process, in which coal is combusted in pure oxygen instead of air. Separation of oxygen from air is energy intensive, and so the energy penalty is still significant. Modification of existing PC plants to use pure oxygen would be substantial as flame temperatures are higher, requiring replacement of many of the basic components of the plant, and so this is unlikely to be a useful strategy for retrofit applications. However, the few demonstration plants that exist today show some promise, largely due to their high thermal efficiency. Whether they will compete economically remains a question.[11]

Another precombustion process for carbon capture is integrated gasification combined cycle (IGCC) technology. Through a process called gasification, coal is heated and partially combusted in pure oxygen to make a mixture of carbon monoxide and hydrogen. The carbon monoxide is then converted to CO_2 through a "shift" reaction with steam, and the CO_2 is collected using separation technologies such as Selexol that use the change in CO_2 affinity from high pressure to low pressure—rather than a temperature change—to regenerate the CO_2 after separation. Much attention has been given to coal gasification as a means for promoting carbon sequestration since studies suggest that the overall costs and the energy penalty are lower for a new plant.[12] However, experience with gasification plants is limited; there are only two such plants in the United States, and neither is equipped to capture carbon (that is, they do not have shift reaction or CO_2 separation capability). More encouragement of coal gasification technology is important to discover whether the promises of lower capture costs can be realized.

10. House and others (2009).
11. Deutch and Moniz (2007).
12. Rubin, Chen, and Rao (2007); Deutch and Moniz (2007).

Overall, there are advantages to each of the different approaches to CO_2 capture, and so each of them is likely to play some role in the U.S. energy system. For example, a power plant located near a petroleum refinery might prefer IGCC technology since the refinery could provide a market for excess hydrogen produced during off-peak hours when electricity prices are low. In addition, some state and local regulators may prefer IGCC technology because of concerns about other pollutants, including sulfur and mercury. In other situations, a new PC plant with postcombustion capture technology might be preferable because the capital costs may be lower. Postcombustion capture technology will likely be a major part of the overall CCS approach because it allows for the reduction of carbon emissions from existing PC plants, at least those with reasonable thermal efficiencies, as discussed above.

Any policies that encourage deployment of capture technologies should be careful to remain neutral in selecting which specific capture technology is best, as competition at commercial scale will be essential to stimulate technological learning and lower costs. A challenge is that the small number of existing IGCC and oxyfuel plants around the world makes it difficult to confidently predict what these plants will cost, how reliable they will be, and whether they will live up to the expectation that their design will make CCS more efficient and less costly. Therefore, some special early incentives for launching these newer technologies will be important over the next decade.

Storage

The technological challenges associated with storage of carbon dioxide from large stationary sources for any individual project are relatively minor compared with those of capture systems, but the scale of the effort is daunting. To achieve the goal of using CCS for all coal combustion in the United States, at current levels of consumption, would require handling more than 2 billion tons of CO_2 a year, roughly double the volume of total U.S. oil consumption when compressed to a liquid state with a density near that of water. Just building the necessary pipeline infrastructure will be a major industrial undertaking, should CCS ever be deployed on such as scale.

The primary scientific question about carbon storage in geological formations concerns the reliability of this approach. Will the CO_2 escape? The good news is that repositories do not have to store CO_2 forever, just long enough to allow the natural carbon cycle to reduce the atmospheric CO_2 to near preindustrial levels. The ocean contains fifty times as much carbon as the atmosphere, mostly in the deep ocean, which has yet to equilibrate with the CO_2 from fossil fuel combustion. Over the timescale of deep ocean mixing, roughly 1,000 to 2,000 years, natural uptake of CO_2 by the ocean combined with dissolution of marine carbonate will absorb 90 percent of the carbon released by human activities. As long as the geological storage of CO_2 can prevent significant leakage over the next few millennia, the natural carbon cycle can handle leakage on longer timescales.

Experience with transport and injection of CO_2 into geological formations comes from decades of work on enhanced oil recovery (EOR) methods in older oil fields. For example, using pipelines built in the 1980s, Kinder Morgan transports CO_2 from natural CO_2 reservoirs in Colorado through a 36-inch-diameter pipe over 300 miles and then injects it into depleted oil reservoirs. EOR demonstrates that CO_2 transport and injection is feasible. Moreover, CO_2 leakage from EOR locations appears to be relatively minor, although careful monitoring and modeling of the fate of CO_2 after injection has not been done in enough detail.[13] There are also some important differences between EOR and CO_2 storage, however, that call into question how useful the experience with EOR will be. First, EOR involves both injection of CO_2 and extraction of fluid—usually a mixture of water, CO_2, and oil (the CO_2 is usually reinjected). The pumping of fluid out of the formation increases the mobility of CO_2, resulting in higher saturation of pore spaces and more effective trapping. Injection of CO_2 into saline aquifers, for example, without concomitant extraction of saline water, may not be analogous to EOR practices in terms of increasing pore pressures and in terms of migration rates of CO_2 in the subsurface. Indeed, displacement of saline water may prove to be a very difficult challenge for terrestrial storage sites as the scale of CO_2 injections grows. Second, EOR may demonstrate that specific oil- and gas-bearing formations that have held hydrocarbons for millions of years can successfully store CO_2 for millennia, but there are not enough depleted oil and gas reservoirs to accommodate the vast volumes of CO_2 if long-term CCS goals are achieved. Overall, EOR does still provide great confidence that CO_2 storage at a massive scale can be accomplished. Moreover, as there is not yet a price on carbon, much less one that would cover the costs of CCS, EOR provides a market for CO_2 over the next decade that can accelerate the deployment of capture technologies.

Most discussion of CO_2 storage focuses on terrestrial injection into formations including deep saline aquifers and old oil and gas fields. In all terrestrial locations at the depth of injection, usually at least one kilometer below the surface, the geothermal gradient means that CO_2 exists as a supercritical fluid with roughly half the density of water. This means that CO_2 can

13. IPCC (2005).

escape if sedimentary formations are compromised by fractures, faults, or old drill holes. The handful of test sites around the world each inject roughly 1 million tons of CO_2 a year, a tiny amount compared to the need for as much as 10 billion tons a year by the middle of the century if most large stationary sources of CO_2 will use CCS to reduce emissions. An important question is whether leakage rates will rise as more and more CO_2 is injected and the reservoirs fill. It seems likely that many geological settings will provide adequate storage, but the data to demonstrate this do not yet exist. A more expansive program aimed at monitoring underground CO_2 injections in a wide variety of geological settings is essential.

Another approach to CO_2 storage is injection offshore into marine sediments, which avoids the hazards of direct ocean injection, including impacts on ocean ecology.[14] If the total depth (both water depth and depth below the sea floor) is greater than 800 meters, then the CO_2 will be in a liquid state with density within 20 percent of seawater (greater than seawater at depths exceeding 3,000 meters). In this case, the mobility of CO_2 would be greatly diminished, yielding essentially a leak-proof repository. This approach may be important for coastal locations, which are far from appropriate sedimentary basins, and may also reduce the extent of expensive monitoring efforts. In addition, offshore storage may be useful to avoid siting pipelines and storage facilities in heavily populated areas.

In terms of capacity, the requirements are vast if most stationary sources will use CCS to reduce emissions. Conservative estimates of reservoir needs over the century are more than 1 trillion tons of CO_2 and might exceed twice that amount. Fortunately, the capacity of deep saline aquifers and marine sediments is more than enough to handle centuries of world CO_2 emissions from burning coal. Matching existing stationary sources of CO_2 with appropriate storage facilities to avoid having to build long pipelines is premature, given that there still is not a single coal-fired power plant in the world that uses CCS technologies, and that the prospects for retrofitting existing plants remain uncertain. However, it appears that the main types of geological storage offer enough options to allow CCS to be deployed in most parts of the United States, either in sedimentary basins on land or in offshore reservoirs in coastal areas in a manner similar to the Sleipner project in the North Sea.

Other forms of CO_2 storage have been proposed, but none has yet shown the promise of simple injection into geological formations, either on land or offshore. Mineralization strategies that would convert CO_2 into carbonate

14. House and others (2006).

minerals appear to be very expensive relative to simple injection, and they have additional challenges associated with moving vast quantities of rock.[15] However, continued research on these and other new approaches is important as CCS goals are likely to require a spectrum of storage strategies for different parts of the country with different geology, state and local regulatory regimes, and levels of public concern.

Moving Forward with CCS Demonstration and Deployment

From the discussion above, it appears that CCS has great potential to reduce CO_2 emissions from stationary sources in the United States (as well as in China and other large emitters); to provide relatively low-cost, low-carbon power; to make use of abundant coal resources in a more environmentally benign fashion; and to make use of some of the existing power plant infrastructure with significant modification but not total reconstruction. The incremental cost of CCS may still be modest when compared to the cost of new nuclear power, new renewable power with storage, or natural gas.

So why are any policy interventions or subsidies needed for CCS projects if CCS is likely to be the most cost-effective way to have low-carbon power? Ultimately, they will not be needed, at least beyond an initial demonstration period, if CCS lives up to its promise. But in the short term, the scientific and technological challenges discussed above must be explored, and enough CCS projects must be operating at a commercial scale so that power developers and investors will have confidence in the technology and in the costs, including those related to liability issues. Through a series of policy recommendations, I discuss the obstacles that must be overcome before there can be widespread adoption of CCS throughout the U.S. energy system. The goals should be to surmount the obstacles to demonstration and initial commercial deployment of CCS systems to learn whether CCS is feasible at a large scale, whether the cost is indeed much lower than alternatives, and whether a proper regulatory framework can be developed using the experience of the initial commercial installations.

Recommendation 1: Provide Federal Subsidies for Commercial-Scale CCS

The U.S. government should provide federal subsidies for ten to twenty commercial-scale CCS projects. These should include different capture technologies (if appropriate) and different strategies for geological storage, and should

---

15. IPCC (2005).

be spread across different regions of the United States to have the biggest impact, both on knowledge gained and also on public perception.

Although CCS should be profitable at some point given a sufficient price on carbon, government assistance is needed in the short term to demonstrate the technology at commercial scale. The price that would be imposed by any of the cap-and-trade legislation currently under discussion in Congress is still well below the level that would cover the cost of sequestration. This may not be a fatal obstacle as investors will anticipate higher prices in the future. But without an adequate price, it is likely that new plants would be built "capture ready" (that is, designed to capture CO_2) but would not actually capture CO_2 and store it in geological repositories. Another problem is that the price of CO_2 could be volatile under a cap-and-trade regime, and this may discourage investment in large capital projects like CCS that depend on a high carbon price.

It should be noted that in several parts of the country, including the Northeast, California, and even parts of the Midwest, it is extremely difficult, if not impossible, to obtain a permit to build a new coal-fired power plant, not because of a price on carbon but simply because local regulators are unlikely to allow any new power plant with high CO_2 emissions. In these markets, a new coal-fired power plant with CCS may actually be profitable today, particularly in the regions dominated by natural gas–generated electricity that have very high prices. But even where CCS is commercially viable today, it is difficult to get investors to assume the technology risks, as well as the risks associated with legal and regulatory issues, including postinjection liability.

A simple way around these concerns is to encourage ten to twenty CCS projects at commercial scales through a variety of government policies and programs. This would allow for demonstration of CCS at enough locations that we would learn whether leakage was a significant problem in certain places, determine which capture technologies were most efficient, and identify any unforeseen problems or challenges. To accomplish this, grants could be awarded on a competitive basis that would pay for some of the incremental capital costs of building a new power plant with CCS. A competitive bidding program might be an efficient way to distribute such subsidies as long as cost was only one of the factors considered when making awards. The Department of Energy, in its restructured FutureGen program, is essentially doing just that at a smaller scale, although sufficient funding has not yet been allocated to the program.

It is essential that government support for commercial demonstration of CCS would also cover the costs of independent monitoring for these projects, at least during the first several years of operation, because knowledge of how the capture technologies operate and what happens to the CO_2 after injection will be important in setting and revising CCS regulations in the future. Within these projects, it would be important to have a range of technologies and storage strategies included. Some of these grants should support retrofit of existing PC plants with postcombustion capture systems, which may require slightly greater funding. Because the intent of these government investments would be to launch true CCS projects and increase our understanding of how such systems would operate at commercial scales, projects that involve storing the CO_2 through EOR should be ineligible for funding. The size of the grants would vary between regions because of differences in the local cost of electricity as well as the existence in some states of subsidy programs for low-carbon electricity that would apply to CCS. Grants would likely need to be higher in coal-intensive regions that currently have very low electricity prices. Awards would not necessarily have to cover the entire additional cost of CCS as these projects may have additional factors that make them more economical, such as accelerated permitting and state subsidies for low-carbon energy. Additional support could come in the form of tax credits that would depend on some minimum fraction of CO_2 captured and stored (for example, 80 percent), or loan guarantees that would reduce the risk to investors in newer technologies including IGCC. All these forms of support could be tied to a carbon price so that the subsidies would diminish if a cap-and-trade bill were passed and these projects were able to benefit from a national carbon price.

Recommendation 2: Create New Federal Laws and Regulations

New federal laws and regulatory policies should be created that would make it easier for operators of power plants and CO_2 storage facilities to understand their liability and to know what environmental regulations will be applied to CCS projects.

The scientific and technical issues discussed above will be confronted over the next decade as the first commercial-scale CCS installations are constructed. However, none of the technical concerns is significant enough to suggest that carbon sequestration cannot be done. But to do it safely and effectively, and to give investors confidence in the approach, a new regulatory regime will be required to address key issues concerning the operation of storage facilities and postinjection liability. These issues are not the focus of this chapter, but they are crucial in moving forward with widespread adoption of CCS technology. Key questions include: Who will certify a storage site as appropriate? How will the capacity be determined? Who will be responsible if CO_2 leaks? Who will safeguard against cheating and how? It is clear that state and federal governments need to play some role in CO_2 storage, just as they do in other forms of waste disposal, but the exact operational details remain murky, which discourages industry from investing in sequestration efforts. However, some agencies have taken a rather narrow view of the CCS challenge. For example, the recent draft of a U.S. Environmental Protection Agency (EPA) rule on CCS only considers possible impacts on drinking water and not the myriad other issues associated with leakage, migration of CO_2 in the subsurface across property lines, or impacts on subsurface processes including earthquake hazards. The EPA rule gives no indication whether the EPA will inspect storage facilities or what the permitting process will be. A broader discussion of a new regulatory regime for disposing of CO_2 is urgently needed.

One key issue that may be decided in the courts is whether CO_2 will be classified as waste, which would make CO_2 disposal subject to many preexisting environmental regulations that were never intended to regulate CO_2 emissions. If the courts rule that CO_2 in CCS applications is indeed waste, then new legislation will be required to specify which rules and regulations apply to CO_2 disposal.

Recommendation 3: Accelerate Permitting and Support Pipeline Installation

The U.S. government should use its authority to encourage state and local governments to accelerate permitting processes for CCS projects. In addition, governments may need to consider supporting pipeline installations for CO_2 transport through grants, tax credits, and rules that will make it easier to site such essential infrastructure.

One of the major obstacles for any new power plant or large industrial construction project is the permitting process, which may involve federal, state, and local regulators. These processes may take several years, which can drive up the price of the projects and may scare away investors. In order to accelerate construction of the first generation of CCS projects in the United States, streamlining of the permitting process will be extremely important and may also drive down the costs. Another local regulatory challenge is the high cost of building pipelines, especially in heavily settled areas where power demand is high. In this case, federal subsidies for CO_2 pipeline construction can encourage CCS, particularly for retrofit of higher-efficiency PC plants (supercritical and ultra-supercritical) with postcombustion capture systems that may be distant from appropriate storage facilities. Siting of pipelines may need to be encouraged via legislation similar to that which allows the Department of Energy to permit transmission lines without consent of state regulators by declaring them in the "national interest."

Recommendation 4: Plan and Provide Assistance for Decommissioning and Retrofitting

If the initial commercialization of CCS is effective, the long-term goal for U.S. energy policy should be the adoption of CCS for all large stationary sources of CO_2 (for example, those emitting more than 1 million tons per year) by the middle of the century or sooner. To accomplish this will require a strategy for shutting down older, low-efficiency plants for which CCS is not a good option. Retrofitting existing plants—not just coal plants—where CCS does make sense may require additional federal assistance beyond the support for demonstration projects as discussed above.

The average age of U.S. electricity-generating plants greater than 500 MW is thirty-two years, and the average age of coal plants is even greater, so most of the capital costs for these plants have been paid for. For these plants, the cost of electricity is primarily a function of the operating costs, including labor, and the cost of the fuel. This is one of the reasons why electricity is so inexpensive in states that have large numbers of older coal plants. This is not true for new power plants, especially PC plants for which capital can contribute as much as half of the total cost of generation. Older plants are also more likely to have low thermal efficiencies and are therefore less likely to be good targets for CCS retrofits, as mentioned earlier. The challenge is that, depending on the cost of construction of new power plants, it may remain profitable to operate these older coal-fired power plants, even with a moderate price on carbon established through a cap-and-trade regime. Rather than simply allowing the price of carbon to rise until such plants are no longer profitable, which may lead to a higher price than what the political system will tolerate, it may be more efficient to encourage the deployment of CCS for plants for which the energy penalty is not too high, and force the decommission of older plants that are incapable of CCS retrofit because of too high an energy penalty, inappropriate design, or a location that is too difficult to match with a storage facility. This will be a political challenge as there are specific regions, such as the Ohio River Valley, that will be affected the most by such policies. Federal funding to subsidize new power projects in these regions may be important to gain support for an overall carbon reduction strategy.

Summary

Carbon capture from stationary sources of fossil fuel combustion and storage in geological formations is an essential component of any comprehensive plan to reduce carbon dioxide emissions in the United States. Scientific and engineering challenges remain, in particular, how to capture and store carbon dioxide from existing power plants that were not designed with this in mind, but none is serious enough to suggest that CCS will not work at the scale required to offset billions of tons of carbon dioxide emissions per year. Widespread commercial deployment of such technologies will require new policies and public investments beyond a price on carbon in order to establish this technology and gain the confidence of investors and regulators. As part of a new national energy policy, a high priority should be placed on accelerating the installation of CCS systems for ten to twenty commercial-scale power plants around the country with different capture technologies, different storage strategies, and extensive monitoring efforts in order to learn what works and what does not, and to gain experience regarding different aspects of CCS systems. If such efforts are successful, and if there is a price on carbon as well as regulatory pressures on carbon emissions, it is reasonable to expect that CCS would become standard for all new coal-fired power plants and many other large stationary sources of CO_2 without additional public funding in the medium term. Additional government policies or subsidies will be needed to add CCS systems to many existing stationary sources, as it is unlikely that a price on carbon alone will be sufficient. For many existing power plants, CCS will not be feasible, either due to engineering restrictions or to the low efficiency of older power plants. In the end, achieving carbon reduction goals will require that these plants be shut down. The ultimate goal of a national CCS policy, if the initial deployment is successful, should be a requirement that such technologies be used for all stationary sources of CO_2 emitting more than 1 million tons per year.

References

Anderson, Soren, and Richard Newell. 2004. "Prospects for Carbon Capture and Storage Technologies." *Annual Review of Environment and Resources* 29:109–42.

Bird, L., and M. Kaiser. 2007. *Trends in Utility Green Pricing Programs (2006)*. NREL/TP-640- 42287. Golden, Colo.: National Renewable Energy Laboratory.

Bohm, Mark C., and others. 2007. "Capture-Ready Coal Plants—Options, Technologies and Economics." *International Journal of Greenhouse Gas Control* 1, no. 1: 113–20.

Deutch, John and Ernest Moniz. 2003. *The Future of Nuclear Power: An Interdisciplinary MIT Study*. Massachusetts Institute of Technology.

———. 2007. *The Future of Coal: An Interdisciplinary Assessment*. Massachusetts Institute of Technology.

Gallagher, Paul, and others. 2006. "The International Competitiveness of the U.S. Corn-Ethanol Industry: A Comparison with Sugar-Ethanol Processing in Brazil." *Agribusiness* 22, no.1: 109–34.

House, Kurt Z., Daniel P. Schrag, and others. 2006. "Permanent Carbon Dioxide Storage in Deep-Sea Sediments." *Proceedings of the National Academy of Sciences* 203, no. 33: 12291–95.

House, Kurt Z., Charles F. Harvey, and others. 2009. "The Energy Penalty of Post-Combustion CO_2 Capture and Storage and Its Implications for Retrofitting the U.S. Installed Base." *Energy and Environmental Science* DOI. 1039.

Intergovernmental Panel on Climate Change (IPCC). Working Group III. 2005. *IPCC Special Report on Carbon Dioxide Capture and Storage*. Cambridge University Press.

McCoy, Sean T., and Edward S. Rubin. 2005. "Models of CO_2 Transport and Storage Costs and Their Importance in CCS Cost Estimates." Paper prepared for the Fourth Annual Conference on Carbon Capture and Sequestration, Alexandria, Va., May 2–5.

Rubin Edward S., Chao Chen, and Anand B. Rao. 2007. "Cost and Performance of Fossil Fuel Power Plants with CO_2 Capture and Storage." *Energy Policy* 35 (March): 4444–54.

United Nations Foundation. Expert Group on Energy Efficiency. 2007. *Realizing the Potential of Energy Efficiency: Targets, Policies, and Measures for G8 Countries*. Washington.

four
Oil Security and the Transportation Sector

Henry Lee

In the aftermath of the Iraq war, the emergence of China as an economic power, and a summer of $4.00 a gallon gasoline, oil security reemerged in 2008 as a major policy issue. Calls for energy independence, increased investments in U.S. oil supplies, and government interventions to reduce demand and promote renewable energy alternatives echoed through the halls of Congress and from the rostrums of candidates for elected office. By fall, however, oil was in free fall, having dropped by $80 per barrel between August and mid-November. While the urgency surrounding the oil security issue dissipated with the price, many experts expect the respite will be short lived, as high oil prices will return when global economic conditions improve. At $40 per barrel, supply investments will decrease, and eventually these lower prices will stimulate demand, setting the stage for another cycle of high prices.

What exactly is the oil security problem? To many it continues to be the threat of a politically induced disruption in the flow of oil into the market, reminiscent of the Arab embargo in the aftermath of the 1973 Yom Kippur War. To others it is linked to a more chronic condition manifested in the micro- and macroeconomic consequences of highly volatile oil prices, expanding trade deficits, and slowing economic growth. To some environmentalists, it is related to emissions from fossil fuel sources, primarily in the transportation sector, and their contribution to greenhouse gas emissions and the threat of global climate change. To poor, developing countries, oil security is the ability (or in many cases, the inability) to obtain kerosene or liquefied petroleum gas for basic needs such as cooking or diesel fuel to transport harvests to local markets. Finally, to those engaged in protecting the United States, either diplomatically or militarily, the oil security problem is the flow of U.S. dollars to certain oil-producing countries who undermine U.S. interests in various parts of the world.

This chapter proposes to answer five fundamental questions: What exactly is the oil security problem, and how serious is it going forward? Why has it emerged at this point in time, and why has it been so difficult for the U.S. government to take the actions needed to mitigate it? Finally, what alternative policies are likely to be effective as the United States attempts to improve its oil security in the future?

The first section of the chapter assesses the oil security problem and attempts to define the factors that will shape it going forward. The second explores the reasons why the United States has been unable to act before this crisis hit and why it is so difficult for it to design meaningful solutions. The final section provides some thoughts on next steps.

The Nature of the Problem

In November 1973, President Nixon called on the United States to "meet its own energy needs without depending on any foreign sources."[1] His focus, as well as that of Senator Henry Jackson, a Democrat from Washington state and the chair of the U.S. Senate Committee on Commerce, was more on the potential of foreign sources to interdict supplies than on the economic impacts of that interdiction. Jackson advocated the establishment of a Strategic Petroleum Reserve (SPR) to supply the U.S. military in times of national emergency. Jackson was responding to the Department of Defense's claim that it needed sufficient oil supplies to be able to support a ground war on two fronts.

After the Iranian crisis of 1979–81, it became evident that the U.S. military's needs paled beside those of the economy, and the SPR was more suited to provide surge capacity to reduce the macroeconomic impacts of disruptions in oil markets. In fact, in almost all U.S. military engagements since 1970, oil supplies were drawn from sources in proximity to the fighting, not from reserves in the United States. The metric for measuring oil security became the macroeconomic impacts of a disruption, as opposed to the physical shortage of supply. In the

---

1. Hakes (2008, p. 26).

ensuing decades, the definition of the problem continued to evolve. By 2008 "energy security" served as an umbrella term under which one could list a number of problems, some more realistic and significant than others.

Reducing Imports

The United States is not its own oil market, but rather part of a larger international market. If one or more oil-producing countries curtail oil shipments to a specific nation, oil is rerouted from other parts of the world to ensure that the embargoed country is supplied. Thus, for an embargo to be even marginally effective, it must be accompanied by a cutback in production, removing oil from the world marketplace and catalyzing higher prices in every part of the globe. Say, hypothetically, that the United States imported zero oil and that there was an embargo on another member of the Organization for Economic Cooperation and Development (OECD), along with a cutback in oil production of an equal amount. Oil prices would increase in Europe and Japan, and there would be a strong incentive for U.S. companies to export some of their oil. However, to make this scenario more realistic, assume that there were government prohibitions on the export of American oil. Europe and Japan are key allies and among our principal trading partners. Their governments would inevitably ask the United States for help. They would explain that unless the United States "shared" some of its domestic oil, the macroeconomic impacts of the disruption would force a drastic reduction in the purchase of U.S. imports. It would be contrary to U.S. economic interests to ignore these pleas, and inevitably oil would flow from the United States to our allies, prices in the United States would rise, and prices in Japan and Europe would fall until once again there was one world oil price.

The bottom line is that no major country or region can politically separate itself from the rest of the world. Oil independence is neither affordable nor desirable. The term may resonate in the midst of political campaigns, as candidates strive to illustrate their concerns about high gasoline prices, but it is not realistic in the globalized world of the twenty-first century.

Funding Our Enemies

In the post–September 11 world, the United States has become increasingly concerned about the threat of terrorism and the spread of anti-U.S. policies in certain oil-producing countries. Iran may be using oil monies to finance the development of nuclear weapons that will further destabilize the Middle East and threaten the future security of Israel. Venezuela uses oil revenues to foster hostility to the Washington Consensus and encourages countries such as Argentina, Bolivia, Ecuador, and Nicaragua to oppose U.S. policies in the region. James Woolsey, former director of the Central Intelligence Agency, pointed out in congressional testimony that over a thirty-year period, the Saudi Arabian government funneled between $85 billion to $90 billion to schools and other institutions to proselytize the beliefs of the Wahhabi fundamentalists, often propagating hatred towards Americans and American interests.[2]

One cannot deny that there are oil-producing countries—and individuals within those countries—who have diverted oil revenues for activities hostile to U.S. interests. However, reducing the volume of imports is not going to reduce these activities. In the first five months of 2008, Saudi Arabia sold an average of 1.6 million barrels of oil per day to the United States. If you annualize this figure and multiply it by the average world price of $97 per barrel (in 2007 dollars), the United States paid Saudi Arabia approximately $56 billion.

Saudi production costs are among the lowest in the world. Thus, if the United States stopped buying oil from that country, the oil would still be sold to nations such as China, Japan, or India, and the Saudis would still receive the revenue, which it could allocate for whatever purposes it saw fit. There is no oil policy that will reduce the flow of revenues to the point that oil-producing countries can no longer afford to fund activities that the United States condemns. The problem described by Woolsey and others is very real. Iran with a nuclear bomb, Chavez's flirtations with Russia, and the continuing efforts by individuals in some Gulf countries to turn an entire generation of Muslims against the United States deserve attention. However, if the United States wants to stop these activities, it will have to rely on levers that have more to do with diplomatic and national security policy, since oil policy by itself is unlikely to deter America's enemies.

High Prices

In 2007–08 the energy security problem was often described by elected officials as high oil prices. This definition implies that the lack of a strong energy security policy was responsible for the high gasoline prices. Advocates of this thesis argue that in the late 1990s and first few years of the present decade, members of the Organization of the Petroleum Exporting Countries (OPEC) deliberately withheld investment in new production capacity so that oil prices

2. See U.S. House of Representatives (2005).

would skyrocket.[3] While it is true that OPEC members have tried to influence the short-term oil market by withholding production, it is less obvious that the same is true for longer-term investments in new facilities. Starting from the time investments are made, it takes at least eight to ten years for a new oil production platform to come on board. Eight years ago the price of crude oil was in the vicinity of $12 per barrel. No one was investing in new supplies, including U.S. oil companies and other non-OPEC producers. Therefore, it is not logical to expect Saudi Arabia to have had the foresight in 2000 to predict $140 per barrel oil prices in 2008 and have made investments accordingly.

The final fallacy surrounding the public impression of energy independence is that its achievement will lower oil product prices. For this to be true, one would have to believe that consumers and suppliers are willingly forgoing the purchase of less expensive substitutes to buy large amounts of expensive imported oil because they have an irrational attachment to expensive imported oil. There is no evidence to support such a theory. Even today the few options that might have cheaper costs, such as ethanol from sugarcane, are priced at the market price for oil products (as opposed to their costs), or their use is restricted by host governments. Without a revolutionary technological breakthrough, a policy for greater energy independence is more likely to increase motor fuel prices than lower them.

If the oil security problem is not about lowering gasoline prices, keeping money out of the hands of our enemies, or physical shortfalls, what is it about? There are four legitimate concerns:
    —short-term economic dislocations from sudden increases in oil prices,
    —long-term supply inadequacies,
    —a foreign policy overly constrained by oil considerations, and
    —environmental threats, specifically global climate change.

Macroeconomic Effects of Short-Term Dislocations

The first of these four concerns focuses on the macroeconomic impacts of a sudden withdrawal of oil from the international market. Disruptions can be politically instigated, result from civil unrest, or be caused by natural disasters. The impact is the probability that such an event will occur multiplied by the size of the oil volume disrupted. In a tight market, a small disruption, or even the threat of such a disruption, can significantly increase oil prices, resulting in reduced GDP, increased current account deficits, and greater

3. Gal Luft, interview by Monica Trauzzi, E&ETV, July 31, 2008. See "Luft Calls for Open Fuel Standard to Break Oil Dependency, Promote Competition in Fuel Sector" (www.eenews.net/tv/video_guide/849 [December 2008]).

monetary inflation. The magnitude of the macroeconomic impacts also depends on the ratio of the value of the oil consumed to total GDP. During the 1980s and 1990s, the dollar value of oil expenditures as a share of total GDP dropped to around 1.1 percent—a small fraction of the 8.3 percent reached in 1980.[4] However, by 2008, when oil prices flirted with figures in the $140 per barrel range and GDP growth was depressed, the ratio increased and came close to the 1980 percentage.

There are three options for dealing with this problem. The first is to encourage the private sector to hedge the future price. This is exactly what many large oil-consuming entities and traders did in 2007 and 2008. As a result, the market for futures and other oil-based derivatives mushroomed. Admittedly, some of the participants may have been more focused on gaining rents than on protecting the value of their existing assets. The impact of this latter group on oil prices is unclear, but it was sufficient to trigger several bills in Congress calling for tighter federal regulation.

The second response is through government-owned or -mandated stocks. As of November 2008, the SPR had over 700 million barrels in storage.[5] In the 2005 Energy Policy Act, Congress mandated an increase in the reserve to 1 billion barrels, but funding to carry out this expansion was not appropriated in 2006 and 2007.[6] Theoretically, when an oil disruption occurs and prices skyrocket, oil can be released, dampening the price increases and their subsequent economic shocks. The problem is that for thirty-two years the U.S. government has been unable to arrive at any consensus on the conditions for releasing the oil in the reserve. In the history of the SPR, oil has been released on only two occasions and loaned to the oil industry on ten others.[7] This has led some critics to pronounce the SPR a "mausoleum" and advocate selling off the oil, especially when oil prices are high and the government can make a healthy profit on its past investments. Others argue that it would be a mistake to lose the reserve. Instead, it would be much wiser to design a protocol outlining the conditions under which the oil would be released from the SPR and establish an independent entity, akin to the Federal Reserve, to manage that protocol.

4. Hamilton (2008, p. 19).

5. A number of other countries, including Germany, Japan, and China, have strategic "reserves," although many of them are managed by domestic oil companies. If the United States is able to design a drawdown protocol, the next step is to begin negotiations as to how best to coordinate the implementation of that protocol with its allies.

6. Bamberger (2008).

7. U.S. Department of Energy, "Releasing Crude Oil from the Strategic Petroleum Reserve" (http://fossil.energy.gov/programs/reserves/spr/spr-drawdown.html [December 2008]).

Ironically, the "strategic reserve" that has been used effectively over the past thirty years is the surge capacity maintained by Saudi Arabia, which consists of about 15 percent of their total production capacity. This reserve gives the Saudi government the ability to release an additional 1.5 million barrels per day onto the market. No other country has been willing to pay the cost of maintaining such a reserve, which in 2007 ranged between $45 billion and $55 billion of forgone gross income. One of the most important questions going forward is, how willing are the Saudis to continue to forgo the revenue that they could receive if they sold this oil? In a tight market, its value will continue to rise and so will the opportunity cost of maintaining it. Thus its availability in the future is not ensured, and there is no other oil-producing country prepared to fill this role if the Saudis decide they would rather have the revenue.

The final option is "boots on the ground"—committing the U.S. military to intervene to protect the flow of oil. While there are widely diverging opinions as to whether oil was a factor in George W. Bush's decision to invade Iraq, his father's decision to enter Kuwait and the U.S. interventions in Iran in the 1950s to supplant Prime Minister Mohammad Mossadegh were influenced by the need to protect the flow of oil. If there is a future civil disruption or political upheaval in the Middle East that results in a withdrawal of a significant volume of oil from the market, especially if it occurs in a tight market, there is a strong possibility that the United States will intervene militarily—or at least threaten to do so. The purpose of this intervention will not be to restore the flow of oil to the United States, since this country receives a small percentage from this region, but rather to protect the world oil market and the flow of oil to Asia and Europe.

Long-Term Instability in the Oil Market

In 2007, 86 percent of the world's oil reserves were owned by state oil companies. Tables 4-1 and 4-2 provide a list of the world's top oil producers and clearly show that the multinational oil companies no longer have a significant impact on global production levels.

Instead, state oil companies and their host governments will make the investment decisions that will determine the world's future oil supply. Often these decisions are based on short-term domestic imperatives or longer-term political concerns, as opposed to commercial factors. The profits these companies make are often the largest source of revenue for their host governments (basically their only shareholder). Thus it is not unusual for short-term politics to drive these companies' decisions on whether to reinvest

Table 4-1. Top Ten Oil Companies by World Petroleum Liquids Production

Thousands of barrels per day

Rank 2006	Company	Production	Ownership
1	Saudi Aramco	11,035	State
2	National Iranian Oil Company	4,049	State
3	Pemex	3,710	State
4	Petroleos de Venezuela	2,650	State
5	Kuwait Petroleum Company	2,643	State
6	British Petroleum	2,562	Private
7	ExxonMobil	2,523	Private
8	PetroChina	2,270	State
9	Shell	2.093	Private
10	Sonotrach	1,934	State

Source: Pirog (2007, p. 4), citing Energy Intelligence Research, "The Energy Intelligence Top 100: Ranking the World's Oil Companies," 2007 edition.

profits or divert them to meet immediate budgetary needs. The recent investment patterns of national oil companies, such as those in Venezuela, Iran, and Nigeria, are quite different from those of private multinational oil companies: shareholders of the former have political agendas that often leave the companies with insufficient funds to meet their long-term production goals.

While the oil markets of the 1980s and 1990s were far from free, the investment decisions critical to future supply flows will become even more politicized. In the period from 1980 to 1992, non-OPEC production increased 82 percent, or slightly less than four times the growth in OPEC production. Demand increased at a growth rate half as much as supply, creating significant surpluses and low prices. But from 1992 to 2007, non-OPEC production only

Table 4-2. Top Ten Oil Companies by Global Liquid Reserves

Millions of barrels

Rank 2006	Company	Reserves	Ownership
1	Saudi Aramco	264,000	State
2	National Iranian Oil Company	137,500	State
3	Iraq National Oil Company	115,000	State
4	Kuwait Petroleum Company	101,500	State
5	Petroleos de Venezuela	79,700	State
6	Abu Dhabi National Oil Company	56,920	State
7	Libya National Oil Company	33,235	State
8	Nigeria National Petroleum Company	21,540	State
9	Lukoil	16,114	Private
10	Qatar Petroleum	15,200	State

Source: Pirog (2007, p. 3), citing Energy Intelligence Research, "The Energy Intelligence Top 100."

increased 18.4 percent while OPEC production increased 31 percent. This trend is likely to continue as the world increasingly relies on OPEC oil, specifically from the Middle East, which continues to be one of the world's most politically unstable regions. Given this decline in the growth level of non-OPEC oil supplies and the decline in the role of private multinational oil companies, future oil supplies are not likely to meet forecasted demand levels. As a result, prices will be more volatile and, if measured over a decade, significantly higher than those experienced over the last twenty years. This trend will be exacerbated by the decline in existing oil fields. Newer fields are further offshore or found in more fractured geological formations, and the crude is heavier in quality. In each of these cases, the oil will be more expensive to produce and refine.

As a result of the politicization of oil resources and the potential underinvestment in new production, consuming countries will find themselves paying more to import oil, which will affect their current accounts and their economic growth levels. Furthermore, inadequate investment will increase the volatility of future oil prices still further. This may be a greater problem for consumers and domestic investors than higher prices per se, since greater volatility will create greater demand risk. Thus, at the margin, the level of investment in either new supplies or alternative energy options will be reduced over what it might have been in a more stable market.

Since 2004 there has been considerable debate about whether or not world oil supplies are close to their physical peak.[8] This thesis is quite controversial, with some experts arguing that world oil supplies will soon decline and others arguing that the decline is still several decades away. While there may or may not be a physical decline in world oil supplies in the next ten years, there is a high probability that there will be a geopolitical peak resulting from the lack of investment in new supplies by oil-producing countries. It is this geopolitical peak, as opposed to the physical peak, that should be of immediate concern to importing countries.

Impact on Foreign Policy

If the United States and the industrialized world were not as reliant on the flow of oil from the Middle East, would U.S. policy toward this region be the same as it has been over the past twenty years—years in which U.S. troops have been sent to Lebanon, Kuwait, Saudi Arabia, and Iraq? If the world becomes more dependent on Middle East oil in the future, will oil security

---

8. See Simmons (2005).

have a greater impact on the options put before U.S. policymakers in the next few decades? Will the United States embrace foreign policy decisions that would otherwise be an anathema, simply because the industrialized countries are so dependent on oil?

There are several difficulties with estimating this "impact." First, it is very hard to measure. How large a factor is oil in shaping U.S. policy? Consensus on the answer to this question is very elusive. Second, if the patterns in the flows of internationally traded oil continue, almost 80 percent of Middle East oil will be going to Asia by 2020. How will this reality affect U.S. policy toward the Middle East? The United States imported 13.2 million barrels per day of product and crude in the first half of 2008, but only 2.4 million barrels or 18 percent was from the Middle East.[9] If Americans are concerned about the influence of imports on U.S. foreign policy, perhaps they should be more concerned about imports from non-OPEC countries, which constitute 52 percent of imports, or those from Canada and Mexico (26 percent).[10]

It would be misleading, however, to deny that maintaining the flow of oil from the Middle East has influenced U.S. foreign policy decisions for close to a half century. Unfortunately, it is equally true that small or moderate reductions in oil consumption are unlikely to change this reality.

Environmental Issues and Global Climate Change

If the Intergovernmental Panel on Climate Change (IPCC) is correct, the world will face larger increases in global average temperatures over the next century than during any period in the last 10,000 years. The impacts and their timing are uncertain, but there is a high probability that unless the nations of the world agree to take significant mitigation action, the world will experience increases in sea levels, the melting of glaciers and portions of the Arctic ice cap, more violent storms, and substantial changes in precipitation. As discussed in an earlier chapter, these in turn will affect demographic trends, land use patterns, water supplies, and regional ecologies. One would be hard pressed to argue that under most of the IPCC scenarios, these changes will not have a major effect on international relations and U.S. foreign policy. For example, climate-induced famine could vastly increase the number of environmental refugees, who in turn will attempt to cross national borders, creating civil instability.

9. Percentages calculated from the *Monthly Energy Review* (U.S. Energy Information Administration 2008), and from U.S. Energy Information Administration, "Petroleum" (www.eia.doe.gov/oil_gas/petroleum/info_glance/petroleum.html [September 2008]).

10. Ibid.

Future secretaries of state will spend an increasing amount of their time negotiating responses to the threat of climate change and designing polices to adapt to the subsequent impacts, which will become larger and more costly in future decades.

Why Has the United States Failed to Address These Problems?

There are many reasons why the United States has been unable or unwilling to respond to these "energy security" problems, but four stand out. First, until 2007 and 2008, oil security concerns were small by almost any measure. The general consensus was that the status quo would continue into the future. Second, there is strong political resistance to any policy that increases the price of oil, and specifically gasoline. Third, there is a belief held by many Americans that new, inexpensive technologies are on the verge of commercialization and that these will resolve both the energy security and climate change threats. The proponents of this view argue that the reason this has not happened is that the United States has lacked the political will to take the strong actions required, but if it elects different officials, the barriers to these technologies will disappear. Fourth, the two biggest threats—lack of investment by oil-producing countries in new supplies in the face of growing worldwide demand and the threat of climate change—require international coordination and cooperation. Mechanisms to facilitate such actions are difficult to achieve and often unable to overcome national self-interest. Finally in 2007–08, some elected officials raised a fifth reason: the country's inability to drill for new domestic oil reserves due to government moratoriums, but as I will show, this hypothesis is much weaker than the other four.

Size of the Problem

Figure 4-1 shows the price of oil from 1980 to 2007. Note that for almost eighteen years, the price remained relatively constant and, relative to today's prices, quite low. Forecasts published in the period 1995–2001 assumed that the prices of the previous ten years would continue into the future.

What was not fully appreciated was how low commodity prices and increased per capita GDP would impact oil demand. It is very difficult to convince investors, and even harder to convince banks, that the future may look very different from the past.[11]

---

11. There is a large literature on this phenomenon in heuristics, anchoring expectations, and the like. See Taleb (2007).

Figure 4-1. World Crude Oil Prices, 1980–2007[a]

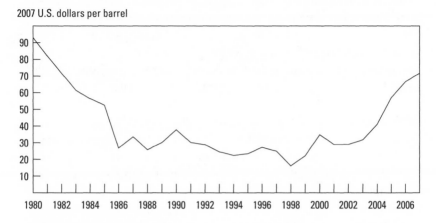

2007 U.S. dollars per barrel

Source: BP (2008, p. 22).
a. Prices for 1980 to 1983 are Arabian Light posted at Ras Tanura; 1984 to 2007 prices are Brent dated.

Many consumers, seeing continuing low gasoline prices and enjoying annual increases in income, substituted new and larger SUVs for their family cars. In 1985 there were 127.8 million cars and 37.2 million light-duty trucks and vans registered in the United States. Twenty years later, the number of cars had increased slightly to 136.5 million, but the number of light-duty trucks and vans exploded to 95.3 million.[12] Miles traveled by all passenger vehicles increased 68 percent during this twenty-year period.[13] Given the low gasoline prices and increased personal income, this behavior made sense. The market signals for the eighteen-year period ending in 2002 clearly influenced consumers to consume more oil and suppliers to invest less. In other words, consumers and suppliers behaved rationally in the face of the low prices they saw in the marketplace.

Policy analysts have various tools to measure the energy problem. Demand elasticity measures the sensitivity of changes in consumers' purchases of gasoline to changes in its price. When prices increase, to what extent do these increases impact consumer demand? A second analytical tool—the energy security premium—measures the externality costs to the U.S. economy from dependence on foreign oil. In each case, the results show that the problem of increasing reliance on imported oil was comparatively small for much of the last twenty-five years.

12. Davis and Diegel (2007, pp. 4.2–4.3).
13. Ibid., p. 3.9.

Short-term elasticity of demand, which measures the response in demand for gasoline to a change in price, dropped precipitously from –0.21 to –0.34 in the 1975–80 time frame to –0.034 to –0.077 in the 2001–06 period.[14] In the latter period, gasoline price increases had very little impact in the short run on vehicle miles driven or on the vehicle purchasing decisions of the American consumer. Furthermore, existing demographic patterns, especially those around major urban areas, do not provide many alternatives to motorists, who have few choices to travel to work, the shopping center, or their children's soccer game.

Paul Leiby at Oak Ridge National Laboratory has been calculating the energy security premium for almost two decades. The premium is the hidden costs or externalities that consumers do not see when they purchase a gallon of gasoline. Leiby asserts that there are two categories of externalities: the macroeconomic costs, such as the costs of reallocating economic resources, GDP growth, and inflation; and the monopsony premium. The latter represents the upward pressure on world oil price that results from U.S. consumption plus the direct costs of importing oil. The United States is such a large purchaser of oil that its purchasing practices affect the world market price. If the United States imported 10 million barrels and reduced its imports by 1 million barrels, the world price would fall. Say that the price of oil is $100 per barrel so that the daily total imported oil bill is $1 billion. If the reduction in U.S. imports causes the world price to fall 1 dollar per barrel to $99, the daily U.S. import bill would decrease to $891 million, or a decrease of $109 million (prices are lower and the volume imported is lower). While the world price would decrease by only $1 per barrel, the resulting decrease in oil purchase payments of $109 million per day is equivalent to a benefit of $109 per barrel of imports reduced or $10 per barrel more than the new world price of $99.[15] The monopsony premium would be $10 per barrel (from $109 to $99), which represents the incremental external benefit to U.S. society as a whole for avoided import costs beyond the price paid for oil purchases.[16] This benefit arises because reductions in U.S. oil imports put downward pressure on world oil prices.

Leiby calculated that in 2004 the monopsony premium was $8.90 per barrel, and the macroeconomic premium was $4.68 per barrel. The total premium was $13.58 or $0.32 per gallon.[17] In 2004 the price of gasoline was

---

14. Hughes, Knittel, and Sperling (2008).
15. The example described here is taken from Leiby (2007, p. 16).
16. Ibid.
17. Ibid.

about $2.00 per gallon, so if all the energy security externalities were included in the form of a tax, the gasoline price would increase only 16 percent.[18] If the U.S. government had decided to add the oil security premium to every gallon sold in the United States, it would not have had a substantial change in consumer behavior or producer investments. In the summer of 2008, gasoline prices had doubled and the security premium was significantly higher. To be overly simplistic, let us estimate that it doubled to $0.64 per gallon. This figure is still relatively small and, if added to the price of oil, would stimulate small to moderate changes in consumer and supplier behavior.

Resistance to Higher Oil Prices

If the goal is to reduce the use of oil or create the market conditions that attract alternative substitutes, one would want higher prices. Higher prices discourage consumption and induce greater investment in the development of alternatives, but this fundamental reality runs into the political reality that initiatives that increase energy prices are unattractive to politicians who are seeking support at the polls. Voters demand lower prices, which will have the opposite effect—increased oil consumption and reduced incentive to develop alternatives. Government policies that result in higher gasoline prices run the risk of being labeled as tax increases, making it very difficult for such policies to gain political traction.

The problem is exacerbated by the historical evidence, which shows that when oil prices drop, as they did in the fall of 2008, the public's concern about oil security and the energy problem plummets. Ironically, the opposite should occur. When oil prices are in a downward cycle, it would seem to be an optimal time to tax oil supplies and use the revenue to invest in a menu of options to reduce demand and increase the supply from substitutes. However, in periods of lower gasoline prices, Congress has historically shown little interest in tackling energy issues. The emergence of global warming as a looming threat may change this cycle.

The Elusive Technological Silver Bullet

Historically, Americans have an inherent optimism that they can design, develop, and nurture the cutting edge technologies that will overcome any

---

18. In 2004 unleaded regular was $1.88 nominal and $1.72 real, and the average of all grades was $1.92 nominal, $1.76 real. Data taken from Energy Information Administration prices for retail motor gasoline and on-highway diesel fuel. See U.S. Energy Information Administration, "Weekly U.S. All Grades Conventional Retail Gasoline Prices" (http://tonto.eia.doe.gov/dnav/pet/hist/mg_tco_usw.htm [January 2009]).

obstacle or meet any challenge. People point to the Manhattan Project that developed the atomic bomb or the efforts in the Kennedy administration to put a man on the moon. What is forgotten in these two examples is that in each case there was one customer—the U.S. government—and that the customer guaranteed it would cover all cost overruns. The task of designing and commercializing new energy technologies is much more complicated.

Automobile manufacturers have the knowledge and capacity to build cars that travel 50–60 miles on one gallon of gasoline. Yet what they do not know is whether consumers will purchase such vehicles. To commit to a multibillion dollar investment for a new assembly line without knowing that there will be a market for the end product is asking the U.S. auto industry to take major financial risks at a time when it is in poor financial health.

In the 1960s and early 1970s, the United States was entranced by the potential of nuclear power and its promise to produce electricity that would be "too cheap to meter." At the end of the 1970s, the public was divided. Some believed that inexpensive synfuels derived from coal would significantly reduce oil imports, while others argued that solar energy would produce measurable portions of the nation's electricity by 1995. More recently, fuel cells have been trumpeted as a means to meet transportation and residential energy needs more efficiently, and biofuels have been touted as the next transportation fuel of choice. Today we are witnessing an infatuation with natural gas from unconventional sources, such as shale formations, and a renewed interest in wind and solar. In almost every case, the hype surrounding these options has not been met. Production costs, lack of consumer acceptance, unanticipated environmental issues, and lower-priced conventional energy sources have made it difficult for these alternatives to penetrate the market to the extent predicted by their advocates.

There is potential in each of these options, and there are strong reasons to pursue their development, but the experiences of the last three decades should have taught us that it is unlikely that we will find a silver bullet. Yet, like the prospector looking for the elusive lode of gold, the American public is easily seduced by the possibility that if the government has the willingness to provide the funding, there are inventors and entrepreneurs who will develop new energy technologies, thus allowing the nation to avoid policy options that might raise energy prices or government regulations that might force the public to change its behavior or habits.

While technology design, development, and dissemination are critical to enabling the United States to meet its energy challenges, no single technology is likely to achieve this end by itself.

Difficulty in Obtaining International Cooperation

Neither the longer-term supply adequacy problem nor the global climate issue can be addressed without international cooperation. The United States is a monopsony buyer of oil. Therefore, if it were able to reduce its oil imports by 2 million barrels per day, the world price would fall and other consuming countries would free ride on the reduced oil prices caused by the U.S. demand reductions. The lower world oil price would stimulate more demand outside the United States. Furthermore, a 2 million barrel decrease in the world oil consumption scenarios developed by both the Energy Information Administration and the International Energy Agency would be small in the context of *total* world consumption—less than 2.5 percent. Two and one-half percent would have an impact on price, especially in a tight market, but the impact would be far greater if efforts in the United States were complemented by actions in other consuming countries. If all of the OECD countries were able to agree on oil consumption reduction targets, either for their own sake or for carbon emission reduction, the capacity to reduce the world's reliance on Persian Gulf oil would be substantially greater.

There are only three options to reduce oil imports: increase domestic production, substitute alternative fuels for oil, or invest in improved energy efficiency. International cooperation could enhance efforts in the last two options; yet achieving such cooperation is difficult. First, different regions have different energy security concerns. In the United States, the primary concern is imported oil and the world's continued reliance on countries that are potentially politically unstable. Policymakers in Europe are focusing on the security of Russian natural gas and on the need to reduce fossil fuel use to meet greenhouse gas reduction targets. Energy planners in China fear that if there is a political schism between the two countries, the United States will interdict Chinese oil and liquid natural gas shipments on the high seas.

Second, effective coordination often runs up against sovereignty concerns and national self-interest. Efforts to prod consuming countries to coordinate their efforts have not been overwhelmingly successful. The International Energy Agency in Paris has facilitated the flow of information among countries but has been less successful in developing any tangible policy coordination.

The bottom line is that designing international coordination mechanisms that work requires that countries agree on the goals, share similar priorities, and are willing to allow domestic policies and programs to be influenced by international imperatives. The experience to date does not give one much confidence that this situation will change in the foreseeable future.

Will Increased Domestic Drilling Make a Difference?

In the 2008 election campaign, many elected officials argued that the continuing moratorium on exploration for oil on certain federal lands was impeding the ability of the United States to produce more oil, thus increasing the need to import supplies from the Middle East. To the American people, frustrated by ever increasing costs of gasoline, this argument hit a responsive chord. What are the facts?

Producers do not know whether a geological formation contains oil or how much oil it might hold until they actually drill. Knowing the size and type of rock that characterizes the geological formations and drawing on past experience, experts can calculate the probability that oil will be found. The U.S. Mineral Management Service calculates that there is a 95 percent probability that there are 66.6 billion barrels of undiscovered recoverable oil in the United States and a 5 percent chance that this number will be as high as 115.13 billion barrels (table 4-3). Of these, 18 billion barrels of possible resources are located in offshore areas covered by federal moratoriums on drilling. An additional increment located on onshore sites, primarily in western states, is also off-limits.

If these areas were opened up to drilling, the National Petroleum Council estimates it would take approximately eight to twelve years to bring any new production onstream.[19] Furthermore, the most promising offshore blocs are off California where the political opposition to drilling is strongest.

The promise of additional domestic supplies of oil is the proverbial glass of water. To some, that glass is half empty; to others, it is half full. If the United States aggressively increased its domestic oil exploration and discovered another 30 billion barrels of oil reserves, hypothetically it could add another 1.5 million barrels per day of domestic production by 2020. Since existing domestic production would be lower than that of 2008, the total increment might be approximately 1 million barrels per day or 8 percent of imports (a conservative estimate). This would be equal to less than 1 percent of the world's oil production.

Would the United States be better off with another 1 million barrels of domestic oil? Would it improve its balance of trade and, in a tight market, relieve some of the upward pressure on price? The answer is yes. On the other hand, would it significantly improve U.S. energy security and have an impact of gasoline prices that would be noticeable? The answer is probably no. This conclusion does not imply that the United States should not pursue this

19. U.S. National Petroleum Council (2007, chap. 2).

Table 4-3. Undiscovered Recoverable Oil of the Outer Continental Shelf[a]
Billions barrels of oil

Outer continental shelf region	F95	Mean	F5
Alaska	8.66	26.61	55.14
Atlantic	1.12	3.82	7.57
Gulf of Mexico	41.21	44.92	49.11
Pacific	7.55	10.53	13.94
Total U.S.	66.6	85.88	115.13

Source: U.S. Department of the Interior (2006).
a. F95 indicates a 95 percent chance of at least the amount listed; F5 indicates a 5 percent chance of at least the amount listed. Only mean values are additive.

option but rather that both proponents and opponents should maintain realistic expectations about its impact.

What Can Be Done?

Shaping solutions to the oil security problem as defined above is difficult for several reasons. First, the problem has two basic aspects: how to reduce the growth in world oil consumption (and, in the OECD countries, the volume of oil imported), and how to reduce greenhouse gas emissions. While the possible solutions to each of these questions overlap in places, there are conspicuous exceptions. For example, producing liquid fuels from coal and, in some cases, fuels from biomass to substitute for gasoline or diesel oil will reduce imports, but it will also increase greenhouse gas emissions. On the other hand, studies on the costs per unit of carbon emissions reduced have shown that initiatives aimed at the transportation sector are more expensive than those aimed at industry and the power sector.[20] This fact does not imply that the transportation sector is not a major source of carbon dioxide emissions. It accounts for 33 percent of U.S. carbon dioxide emissions, with motor vehicles, excluding passenger cars and trucks, accounting for 20 percent.[21] But if the only problem was greenhouse gas emissions, policymakers might be inclined to focus less on mobile sources and more on stationary facilities, while if the problem was focused on reducing oil imports, policymakers would reverse their emphasis.

20. McKinsey and Company, "Reducing U.S. Greenhouse Gas Emissions: How Much at What Costs. U.S. Greenhouse Gas Abatement Mapping Initiative," December 2007 (www.mckinsey.com/clientservice/ccsi/pdf/US_ghg_final_report.pdf).
21. Gallagher and others (2007).

Table 4-4. Oil Consumption in U.S. Transportation Sector, 2006

Units as indicated

Mode of transportation	Thousands of barrels per day	Percent of total
Cars	4,891.3	36
Light trucks	3,957.1	29.1
Motorcycles	14.4	0.1
Buses	93.2	0.7
Heavy trucks	2,473.5	18.2
Air	1,208.3	8.9
Water	663.9	4.9
Pipeline	5.3	0
Rail	285.4	2.1
Total	13,592.4	100.0

Source: Davis and Diegel (2007, table 1.14).

Second, it will be very difficult to change the basic demographic and physical infrastructure that now exists. There are two components to this infrastructure: the first affects how the country uses oil, and the second affects the way in which oil products are distributed and sold.

Table 4-4 shows that in 2006 approximately 68 percent of U.S. oil consumption occurred in the transportation sector, and within that sector almost two-thirds was consumed by automobiles and light trucks. There are three regulatory policy options to reduce oil consumption in this sector:

—require that new cars be more efficient (use less gasoline per mile driven);

—require that a greater portion of the energy used to power the vehicle come from sources other than oil, such as electricity, natural gas, or biofuels; and

—require that the vehicles drive fewer miles.

Studies at Harvard University's Belfer Center for Science and International Affairs show that to have a significant effect on oil consumption, policymakers will have to pursue all three.[22] Yet each is fraught with problems.

Fuel-Efficient Vehicles

In 2007 there were 232 million passenger cars in the United States.[23] The average life of a vehicle is estimated at thirteen years, meaning that it will take at least this long for newer, more efficient cars to replace the older, less efficient

22. Ibid.
23. Davis and Diegel (2007).

models.[24] If the more efficient cars have lower performance or are seen as less attractive, the rate of dissemination will be even slower.

In 2007 Congress passed the Energy Independence and Security Act, which increased the Corporate Average Fuel Economy (CAFE) standard from 27.5 miles per gallon for cars and 22.2 miles per gallon for light trucks in 2007 to 35 miles per gallon for all light-duty vehicles (cars and light trucks) by 2022, while expanding the vehicle fleet to which these standards apply.[25] The hope was that the new law would significantly reduce oil consumption. Yet subsequent analysis shows that the income effect, as millions of Americans become wealthier and per capita GDP increases, will swamp the impact of having more efficient cars on the road. The analysis shows almost no reduction in oil imports or carbon emissions in 2030 relative to 2010 base emissions.[26] This is the same phenomenon witnessed with the original CAFE standards. Since these standards were enacted, vehicle fuel consumption in the United States has increased 60 percent, primarily because people bought many larger and more powerful cars and drove more.[27]

Substitute Fuels Face Similar Challenges

The Energy Independence and Security Act requires that 36 billion gallons of renewable fuel be blended into gasoline by 2022. Some of this fuel is projected to come from corn-based ethanol, with the remainder to come from second-generation biofuels, which are projected to rely heavily on cellulosic material. There is rising concern, however, that carbon emissions over the lifecycle of the production chain for some biofuels may not be significantly less than the carbon emissions from gasoline production. In addition, the biomass that provides the feedstocks will require significant amounts of land, during a period when the world's population is continuing to grow. Using land previously dedicated for food will increase food prices, put added pressure on forests, threaten biodiversity, and put pressure on water supplies. Furthermore, if the United States wants to pursue biofuels as an energy option, as opposed to stimulus for American farmers, it should lower its import tariffs and encourage the use of cheaper biofuels produced from sugarcane. These problems are solvable, but not without significant government intervention.

24. U.S. Congress, Joint Economic Committee (2007, p. 5).
25. The estimated 2007 model year light-duty vehicle fleet efficiency (consisting of cars and light trucks) was 20.2 miles per gallon, meaning that the mandated increase in new vehicle efficiency is quite dramatic (U.S. Environmental Protection Agency 2007).
26. Gallagher and others (2007).
27. Ibid., p. 13.

Other potential substitutes face equally daunting challenges. Many people point to the potential of liquids from coal, but the production processes emit too much carbon to be environmentally acceptable and will require substantial investments in technologies to capture and store the carbon emitted. The production of unconventional gas supplies requires substantial amounts of water and faces steep regulatory hurdles. Hydrogen has enormous potential, but it is a secondary energy source, and its production requires large amounts of electricity. Finally, electric vehicles will need to increase their range if they are to attract buyers. To achieve this end, breakthroughs in battery technologies will be needed.

All of these options have potential, but all face major challenges and will not make significant contributions without major efforts by both government and the private sector.

Vehicle Miles Traveled

If car efficiency is improved and a portion of the fuel is produced from oil substitutes, the potential to reduce oil imports and carbon will not be fully met unless the United States can also reduce the amount of vehicle miles traveled (VMT). Figure 4-2 shows that from 1990 to 2006, vehicle miles traveled in the United States increased every year. Growth in VMT was negative in 2007 for the first time in over twenty years. While, as of this writing, the 2008 data are incomplete, July figures indicate that VMT numbers were approximately 3.6 percent lower than in July 2007. When the economy recovers, these figures are likely to increase.

The difficulty in reducing VMT is that it is a source of enormous economic and social benefits. Brian Taylor at UCLA's Institute for Transportation Studies produced an excellent overview of vehicle travel.[28] He argues that travel benefits society by enabling multiple economic transactions and allowing the global integration of labor markets. There is a direct correlation between VMT growth and economic prosperity.

Most American metropolitan regions sprawl across large areas, and people need to drive to their jobs, to the shopping centers, to their children's schools, and to various recreational activities. Taylor points out that trips for errands now exceed trips to work by 2.5 to 1.[29] It will take decades to change the spatial structure of our metropolitan areas. Thus changing travel patterns in the

28. Taylor (2008).
29. Ibid.

Figure 4-2. Total Vehicle Miles Traveled (VMT), 1990–2007

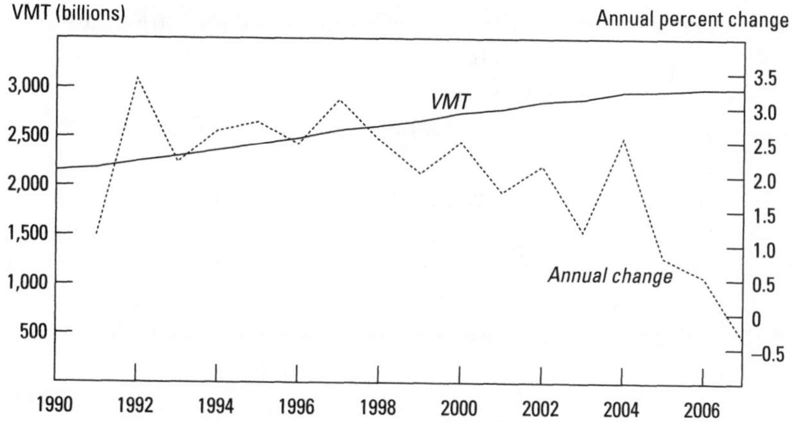

Source: Based on statistics from Research and Innovative Technology Administration, U.S. Department of Transportation.

short term will be very difficult. This conclusion is supported by the recent experience in some European countries that have flirted with $8–$10 per gallon gasoline without major reductions in the number of trips and the distances traveled.[30] While a successful effort to reduce oil imports requires a reduction in VMT, it will not be an easy task to make measurable reductions without incurring stiff consumer resistance and potential economic losses.

Infrastructure

If the country moves to a transportation system that relies on biofuels, compressed natural gas, or electricity, new infrastructure facilities will need to be developed. Plug-in hybrids will need to be recharged during the day, which will mean constructing a network of charging stations or installing outlets in garages and parking areas. One can imagine gasoline stations that have a pump for gas, a pump for ethanol or an ethanol blend, such as E85, and a natural gas line. Not only would each existing station have to be retrofitted, but investments would be needed to transport the fuel to the station. There are many questions, starting with who will take the demand risk of investing in these new energy networks? How much will they cost, and how quickly could they be put in place?

30. European Union Road Federation (2008).

Putting a Price on Oil Consumption and Carbon Is Essential

If the United States is to reduce its use of oil and its level of carbon emissions, it will need to place a price on both imported oil and carbon either by taxing them or via a cap-and-trade program. In a world where politicians are promising lower prices and greater oil independence, this will be an enormous political challenge.

By increasing the prices of imported oil and carbon, the government would send a strong signal to consumers to use less oil and emit fewer grams of carbon. It would also send a signal to the thousands of entrepreneurs who are striving to develop and commercialize substitutes for imported oil. With the earlier example of reducing oil use in passenger vehicles, a higher price on oil will stimulate consumers to purchase more fuel-efficient vehicles, use substitutes for gasoline, and reduce VMT. A recent study by Gallagher and Collantes showed that to gain significant reductions in passenger vehicle carbon emissions, the government would have to impose both an economy-wide carbon tax of $30 per ton of carbon dioxide and a gasoline tax that would start at 10 percent and escalate each year until gasoline prices were slightly under $6.00 per gallon.[31]

Given the political resistance to taxes, or even a cap-and-trade program that meaningfully increased oil prices, one might reasonably conclude that it is unrealistic to expect Congress to take action to raise prices in the face of strong popular opposition. This may be true, but if Congress is serious about reducing oil imports or carbon emissions, or both, and is willing to entertain passage of such an initiative, it should ensure that the revenue to the government is maximized. Under a cap-and-trade program, this would mean auctioning as high a percentage of the allowances as politically possible. A substantial portion of the revenue derived should be used to offset other taxes, preferably those paid by a large ratio of the population, such as Social Security taxes. Several members of Congress would prefer to roll these monies into an "energy development fund" that would support an array of alternative energy options. The nation, however, would be better served if project selection was left to the market and a substantial percentage of the revenue recycled to the taxpayers.

The remainder of the funds should be used to cover transitional costs that fall into two categories: financial support for those who will have difficulty

31. Gallagher and Collantes (2008).

affording the additional costs and funds to cover a portion of the cost of new infrastructure necessary to support emerging substitute fuels. In the former, funds would help low-income oil users bear the burden of higher prices, especially those consumers still reliant on oil for heating. Assistance would also be justified for those industries that are critical to the transition. In the latter, one might include financing for new infrastructure, such as rail lines to move goods, pipelines to move gas, new transmission to move electricity, and distribution networks for biofuels or, sometime in the future, hydrogen.

If Congress is reluctant to support a proposal that increases oil prices, it could consider a variable tax that would be triggered when oil prices reach a certain threshold. For example, say oil prices are $80 per barrel. Congress could set a floor price of $60 per barrel, and if oil prices slipped below $60 to $50 per barrel, a $10 tax would be imposed. If one wanted to maximize the assistance to U.S. oil producers, one could target the tax on oil imports. If several months later the price increased above $60, the tax would disappear. Investors and consumers would know that the price would not fall below $60 and would make investments accordingly. Such an action would significantly reduce the demand risk that causes entrepreneurs to underinvest in substitutes for oil and consumers to forgo purchasing more energy-efficient vehicles. The window of opportunity for a variable oil tax disappeared as world oil prices for late November 2008 slipped below $50 per barrel. However, if and when oil prices go back up again, the variable tax could make significant contributions to creating a more stable investment climate for both investors and consumers.

Many observers are skeptical that Congress would be willing to set a price sufficient to make a difference for either carbon emissions or imported oil. Therefore, they believe that the government should focus on other measures. At the risk of oversimplification, there are three alternatives that seem to dominate the search for options to reduce oil consumption in the transportation sector, other than putting a price on emissions or oil: regulating behavior and requiring the use of specific technologies, taxing the motor vehicle rather than the fuel, and taxing road congestion. These are not mutually exclusive and could be used to complement a carbon price or a tax on imported oil.

Regulation

The California climate program relies on a portfolio of mechanisms, including regulating the use of new, more energy-efficient technologies and requiring multiple sectors of society—industry, commercial, and government—to find ways that they can reduce their carbon emissions. To ease the economic costs, the state has relied more on performance standards than requiring specific technologies or behavioral adjustments. That is, instead of telling the trucking industry to install a specific piece of equipment, it prefers to tell them to reduce their emissions by a certain percent, leaving to them the task of figuring out the most cost-effective alternative. Where feasible, the state also allows companies that face very high carbon abatement costs to purchase emission allowances from firms that have lower costs.

While California's approach is comparatively cost-effective, the real question is: can one realize the goal of reducing oil use through heavy reliance on a portfolio of dozens of regulations? Each performance standard is set based on what is technologically possible at the time the regulation is issued, but these technologies are constantly changing. Thus regulators must either regulate to the technological status quo in a dynamic market or try to predict those technologies that may emerge in the future. Neither path is likely to lead to the optimal mix of technologies and behavioral changes. In the case of the first, the state ends up setting a ceiling that results in emission reductions that are much less than what society would otherwise seek; in the second case, the state is likely to misjudge the direction and scope of technological changes and thus create disincentives to meet an optimal level of emission reduction, or if the projected technologies do not emerge, create a situation whereby the state requires emission reductions that are not feasible or are extremely costly. In the latter instance, government usually has to amend its regulations, and the process of changing the regulations often carries very high transaction costs—both political and economic.

Furthermore, if consumers do not receive effective price signals, many will develop strategies to game the regulations so that their actions meet their individual self-interests. The state is forced into a situation where it is continually adjusting the regulations to close loopholes while trying to enforce multiple regulations to reduce oil use in a low-priced oil market. History tells us that this scenario rarely results in meaningful reductions.

All of these costs would apply whether California or the federal government is pursuing a regulatory approach. In fact, one might argue that California does not have the jurisdictional capacity to rely on a strategy that emphasizes pricing. However, having one state develop a package of regulations to meet the goal of either decreasing carbon emissions or reducing oil imports leads to this question: what are the consequences if the other forty-nine states adopt a similar regulatory approach but design different packages? The transaction costs would be enormous and would make the ongoing problems surrounding boutique gasoline blends a small blip on the economic radar.

Taxing the Vehicle

A second option would be to tax the vehicle. In countries such as Israel, Singapore, and Denmark, taxes on automobiles are so high that cars cost two to three times the going price in the United States. If these taxes were applied so that efficient cars or flex fuel cars were taxed at much lower rates than less efficient models, one could provide incentives for consumers to purchase the former. One could also use this model to create a market for electric cars or plug-in hybrids. These taxes can take the form of levies at the point of sale, or they could be annual feebates, collected in a manner similar to excise taxes.

The problem with this alternative is twofold. First, it only works if the tax on the less fuel-efficient vehicle is very high and a significant number of the conventional models now being sold are considered "fuel efficient." No one would grouse if such a tax were placed on a half-dozen European sports cars that only Hollywood celebrities can afford, but if it were suddenly applied to a Ford Taurus or a $22,000 Chevrolet, consumers would protest. Second, most versions of this tax would only affect the purchase of new cars and thus have the same problem as CAFE standards: no meaningful reduction in miles traveled and no incentive to change fuels. It is possible to design a feebate scheme that would incorporate miles traveled, so that if one bought a Hummer and never drove it, one would pay a lower tax than someone who drove his Hummer 12,000 miles a year. However, most of the currently discussed feebate proposals do not address VMT.

Congestion Fees

A third alternative strategy that receives considerable attention is a plan that relies on congestion fees. Such fees would collect tolls on roads, with higher rates being levied during times of congestion. Given that miles driven have increased dramatically in the last twenty years and very little new infrastructure has been built, congestion around many metropolitan areas has increased dramatically, becoming a major problem. Furthermore, there is a very large transportation infrastructure deficit in the United States; congestion fees would provide a source of badly needed funds.

The problem with this option is that less congestion may result in an increase in VMT. Congestion is a disincentive to drive, and lack of congestion provides the opposite incentive. Thus a program that successfully reduces congestion ultimately may increase the demand for oil imports. Second, congestion fees fall within the jurisdiction of state and local governments, not the federal government. Therefore, one would have to persuade 50–100 different jurisdictions that such fees are merited.

Each of the aforementioned three options has merits, but if the goal is to reduce oil consumption, none will have a significant impact unless it occurs within the context of strong price signals.

Government's Role

In an energy transition driven by higher prices, government will have four roles. The first is easing the financial burdens inherent in the transition, as discussed earlier. The second is establishing environmental goals, setting standards to meet those goals, and then enforcing those standards. The third is providing funds for research and development aimed at finding basic engineering and technological breakthroughs that will increase the availability of supply and demand alternatives. Finally, if international coordination and cooperation are desired, governments must be involved in designing the protocols, treaties, and other informal and formal arrangements.

Reducing the externalities, particularly environmental, inherent in the production and use of non-oil-based transportation fuels will be an enormous challenge and will require government to play a lead role. For example, developing substitutes for oil-based transportation fuels can result in significantly lower carbon emissions than those emitted in the production and use of gasoline and diesel fuel, or it could result in much higher emissions. Government has the opportunity to set the standards that will shape the carbon intensity of new alternatives, such as biofuels, compressed natural gas, or even liquids from coal or biomass.

While the United States has relied on energy efficiency standards to improve the energy productivity of the transportation fleet, Europe and, more recently, California have opted to require vehicles to meet performance standards for carbon emissions. The latter is more focused on a clear goal and provides fewer opportunities to exploit loopholes. Both create competitive distortions that regulators must manage. The political problem in the United States is that automobile efficiency standards are in place and are implemented by the Department of Transportation. Carbon-based performance standards are not in place and would be implemented by the Environmental Protection Agency. Certain constituencies, such as the automobile companies, are more comfortable with the Department of Transportation (with whom they have dealt for three decades) than they are with the Environmental Protection Agency.

While establishing environmental standards and goals is essential, the federal government must assume two additional responsibilities. First, it must provide technical and financial assistance to subnational governments, including state, local, and regional bodies, to reduce environmental externalities and to invest in a new energy infrastructure. Second, it must invest in the research and development of technologies and methods to reduce the environmental footprint of alternative options. For example, increased reliance on biofuels will put significant additional burdens on the use of land. It will do so at the same time that population and per capita income growth will increase food demand. It will also impact, either directly or indirectly, the use of forested lands and will affect the uses of local water supplies. Decisions affecting the uses of land and water supplies are not usually made at the national level but rather at state, local, and regional levels. Coordination between these levels of government has been disjointed. Furthermore, subnational levels of government do not have the information and technical knowledge to make sustainable land use decisions. There is a need to substantially improve the delivery mechanisms to develop and deploy information and knowledge from the federal to subnational levels of government.

Land use decisions require that governments consider energy, agricultural, environmental, and economic ramifications. Yet knowledge of each resides in separate departments at the federal and state levels. This stovepiped allocation of responsibility creates duplication of effort and decisions that reflect the interests and mandates of one agency instead of all the concerned parties. If the United States increases production of crops for fuel and development of unconventional natural gas deposits or alternative electricity options, such as wind or solar energy, integrated land use decisionmaking will grow in importance.

Finally, advancing a strategy for sustainable development of biofuels that meets concerns about availability, cost-effectiveness, greenhouse gas reductions, food production and consumption, and ecosystem protection will be a knowledge-intensive activity.[32] Historically, very little research has been supported in the areas of improving land productivity, protecting multiuse landscapes, and ensuring sustainable usage of declining water supplies. There has

32. Lee, Clark, and Devereaux (2008, p. 13).

been almost no research on developing new crops that would not compete with food production. If the United States is to meet a target of having bio-fuels constitute 20 percent of its transportation fuels, research into these top-ics must expand.

International Coordination

If the goal is to reduce the amount of oil consumed, and there is a single world oil market, then the amount of oil demand must be reduced. Focusing on one oil-consuming country, such as the United States, and ignoring other major oil consumers, such as China or Europe, will be far less effective in reducing world oil demand than a coordinated effort that reduces oil con-sumption in all the major importing countries. A strong and focused pro-gram might be able to reduce U.S. consumption of imported oil by 2.5 mil-lion barrels per day over ten to fifteen years. Assuming that world oil demand is 105 million barrels per day in 2020 and that most of the incremental sup-ply—compared with 2008—will come from the Middle East, such a change in U.S. consumption would reduce the world demand for Middle East oil by 9 percent. If the U.S. figure could be doubled by decreases in demand from other countries, then the percentage reduction would increase to 18 percent. In a tight market, reduction of this magnitude could have a measurable im-pact on the price of oil.

The same argument can be made for carbon reductions, except that in this instance the world has seen what happens when only some countries agree on an international protocol. Not only is the impact on world concentrations of greenhouse gases much smaller, but it creates distortions in patterns of trade and economic growth. If all other factors are held equal, manufacturers in countries that have not committed to reduce carbon emissions have a com-petitive advantage. There is an incentive for emission leakage at the margin, which will induce carbon-emitting companies to relocate to countries that do not require reductions in carbon emissions.

Therefore, greater international coordination—either through bilateral or multilateral agreements—will result in greater reductions in world oil con-sumption than if every country unilaterally pursues reductions.

Other Thoughts

The key to unleashing the creativity, technological expertise, and consumer behavioral adjustments is to price oil consumption so that it internalizes the costs of both the security and environmental externalities. Such a pricing regime will also raise revenues to cover some of the costs inherent in the transition to a less carbon-intensive energy mix and accelerate desired behavioral changes in the transportation sector. However, a tradable permit price of $100 per ton of carbon translates into an increase in the price of gasoline of only $0.26 per gallon—not sufficient to stimulate even a moderate change in either oil consumption or carbon emissions.

Thus one could argue that stronger incentives are needed. The most efficient option would be either a tax on oil use in the transportation sector or a separate cap-and-trade program. In their recent paper, Gallagher and Collantes show that a $10 per ton carbon price plus a 2 percent annual improvement in vehicle efficiency and a 50-cent-per-gallon gasoline tax, escalating at 10 percent per year, would reduce energy consumption in the light-duty vehicle sector by 4.8 quadrillion BTUs, approximately 2.3 million barrels of oil per day, relative to a base case in 2030.[33] The political feasibility of such a package is low, but it provides an idea of what would be required to have a significant impact on oil consumption.

If an economy-wide oil tax or cap-and-trade program proves to be politically unacceptable and Congress decides to promote specific technologies, which technology looks to be most promising? Despite the claims to the contrary by their advocates, there is no way to tell whether plug-in hybrids are superior to diesel technologies or that compressed natural gas has more potential than third-generation biofuels. If government decides to pick winners, it will most likely select the wrong options. However, if government is determined to minimize its reliance on pricing policies and focus on technology development, it should support a wide portfolio of options. As previously mentioned, it is more probable that different technologies, and even different fuels, will emerge in certain regions and under different conditions. For example, plug-in hybrids may make sense in a densely populated urban area while diesel cars may make more sense in less urbanized areas where people drive greater distances. Biofuel blends may make more sense in Missouri than in California. Hence, if Congress decides to go down this route, it should seed a portfolio of options and allow the marketplace to select the winners.

Conclusion

There is a high likelihood that the national oil companies that sit on more than 86 percent of the world's oil reserves will underinvest and that supplies

---

33. Gallagher and Collantes (2008). As a reference, the transportation sector consumed 29 quadrillion BTUs in 2007 (Davis and Diegel 2007, pp. 2–3).

will prove less than the numbers now projected by the economic models used by both the U.S. Energy Information Agency and the International Energy Agency in Paris. The threat of a geopolitical peak is very real. If this occurs, future oil prices will prove to be more volatile, with higher increases on the upside of the price cycle. Monopsony premiums calculated by Leiby in 2004 will prove to be too low, and the energy security premium will be much higher.

Reducing oil imports requires more efficient modes of moving goods and people, use of alternative fuels, and reductions in vehicle miles traveled. While all three will be difficult to achieve, reductions in VMT will prove to be the greatest challenge. If prices for oil products including gasoline decrease, elected officials may celebrate in the short run, but it will be very difficult, if not impossible, to achieve meaningful reductions in consumption in the transportation sector. Despite strong government regulation and subsidies, consumers and producers alike will not change their preferences or their behavior sufficiently to stimulate meaningful reductions in oil imports. Unless there is a clear price signal, the probability that government can rely exclusively on subsidies, grants, and guarantees to stimulate the development of new cost-competitive technologies that will be disseminated throughout the economy in the next decade is remote. To change the odds, one must change the marketplace, and this requires pricing carbon and oil imports at levels that will stimulate changes in both supply and demand. There is, however, an essential dilemma. Implicit in these arguments is that the externality costs of emitting greenhouse gasses and importing oil justify strong government intervention. As discussed earlier, Leiby calculates that the oil import premium is $0.32 per gallon.[34] The highest externality value for climate, contained in the *Stern Review on the Economics of Climate Change*, is $0.75 per gallon.[35] The combined total is $1.07 per gallon, which is significantly less than the amount required to induce measurable reductions in the volume of petroleum consumed in the transportation sector. Therefore, policymakers either have to cap the price increases at a figure considered by many to be inadequate, or they must argue that the externality figures are badly underestimating the true damage functions. There is no right answer to this problem; only future historians will be able to determine whether this generation of Americans chose the correct path.

34. Leiby (2007).
35. Stern (2006).

References

British Petroleum (BP). 2008. *BP Statistical Review of World Energy 2008*. London.

Bamberger, Robert. 2008. "The Strategic Petroleum Reserve: History, Perspectives, and Issues." Report RL33341. Congressional Research Service.

Davis, Stacy C., and Susan W. Diegel. 2007. *Transportation Energy Data Book*, 26th ed. Oak Ridge, Tenn.: Oak Ridge National Laboratory.

European Union Road Federation. 2008. *European Road Statistics 2008*. Brussels.

Gallagher, Kelly Sims, and Gustavo Collantes. 2008. "Analysis of Policies to Reduce Oil Consumption and Greenhouse Gas Emissions from the U.S. Transportation Sector." Discussion Paper 2008-06. Cambridge, Mass.: Belfer Center for Science and International Affairs (June).

Gallagher, Kelly S., and others. 2007. "Policy Options for Reducing Oil Consumption and Greenhouse Gas Emissions from the U.S. Transportation Sector." Energy Technology Innovation Policy Discussion Paper. Harvard Kennedy School (July 27).

Hakes, Jay. 2008. *A Declaration of Energy Independence*. Hoboken, N.J.: John Wiley and Sons.

Hamilton, James D. 2008. "Understanding Crude Oil Prices." Energy Policy and Economics Working Paper 023. University of California Energy Institute (June).

Hughes, Jonathan, Christopher Knittel, and Daniel Sperling. 2008. "Evidence of a Shift in the Short Run Elasticity of Gasoline Demand." *Energy Journal* 29, no. 1: 113–34.

Lee, Henry, William C. Clark, and Charan Devereaux. 2008. "Biofuels and Sustainable Development. Report of an Executive Session on the Grand Challenges of a Sustainability Transition." Harvard University, Center for International Development, Sustainability Science Program.

Leiby, Paul. 2007. "Estimating the Energy Security Benefit Reduced Oil Imports." ORNL/TM 2007/028. Oak Ridge, Tenn.: Oak Ridge National Laboratory.

Pirog, Robert. 2007. "The Role of National Oil Companies in the International Oil Market." Report RL34137. Congressional Research Service.

Simmons, Matthew. 2005. *Twilight in the Desert*. Hoboken, N.J.: John Wiley and Sons.

Stern, Nicholas. 2006. *Stern Review on the Economics of Climate Change*. Cambridge, UK: HM Treasury.

Taleb, Nassim. 2007. *The Black Swan*. New York: Random House.

Taylor, Brian. 2008. "Influencing Vehicle Travel and Transit Use in the Years Ahead." Presentation to Committee for the Study of the Potential Energy Savings and Greenhouse Gas Reductions from Transportation. UCLA Institute of Transportation Studies, Irvine, California, May 1.

U.S. Congress. Joint Economic Committee. 2007. *Money in the Ban—Not in the Tank*. Report. 109 Cong. 2 sess.

U.S. Department of the Interior. Minerals Management Service. 2006. "An Assessment of Undiscovered Technically Recoverable Oil and Gas Resources." MMS Fact Sheet RED 2006-01b (February).

U.S. Energy Information Administration. 2008. *Monthly Energy Review* (July).

U.S. Environmental Protection Agency. 2007. "Light-Duty Automotive Technology and Fuel Economy Trends: 1975 through 2007. Executive Summary." Technical Report EPA420-S-07-001 (September).

U.S. House of Representatives. Committee on Government Reform, Subcommittee on Energy and Resources. 2005. "Testimony of R. James Woolsey, National Commission on Energy Policy." 109 Cong. 1 sess. (April 6).

U.S. National Petroleum Council. 2007. *Hard Truths: Facing the Hard Truths about Energy.* Washington.

five

Policy for Energy Technology Innovation

Laura Diaz Anadon and John P. Holdren

The United States and the world face pressing economic, environmental, and security challenges arising from the energy sector, paramount among them providing the increased quantities of affordable energy needed to meet economic aspirations, limiting the political and economic vulnerabilities of heavy dependence on oil, and reducing the risk of unmanageable disruption of global climate by emissions of carbon dioxide from all fossil fuel burning. Improving the technologies of energy supply and end use is a prerequisite for surmounting these challenges in a timely and cost-effective way.

The United States ought to be the leader of the world in the energy technology innovation that is needed. It has the largest economy, uses the most energy (and within that total the most oil), has made the largest cumulative contribution to the atmospheric buildup of fossil carbon dioxide that is the dominant driver of global climate change, has a large balance of payments stake in competitiveness in the global energy technology market as well as a large stake in the worldwide economic and security benefits of meeting global energy needs in affordable and sustainable ways, and possesses by many measures the most capable scientific and engineering workforce in the world. The actual performance of this country in energy-technology innovation, however, has been falling short by almost every measure: in relation to the need, in relation to the opportunities, in relation to what other countries are doing, and even in the simple-minded but still somewhat instructive measure of investment in energy-technology innovation in absolute terms and as a proportion of GDP, compared to the past.

Current U.S. federal government investments in energy research, development, and demonstration (ERD&D) are about the same in absolute terms as they were twenty-five years ago—and thus less than half as large as twenty-five years ago in relation to GDP. Japan has passed the United States in absolute magnitude of public expenditures for ERD&D and, with correction for purchasing power parity, China may soon do so. Private sector ERD&D expenditures are harder to track, but this much is clear: the pace of improvement of key indicators of energy sector performance (such as the ratios of energy use, oil use, and CO_2 emissions to real GDP) is far short, in the United States and worldwide, of what will be needed to surmount the major challenges; and, in such key sectors as renewable energy and energy end-use efficiency, leadership in innovation has been passing from the United States to Europe and Asia.

Public and private ERD&D expenditures are small in relation to the economic, environmental, and security stakes and even in relation to total national expenditures for energy itself. U.S. federal budget authority for energy-technology research, development, and demonstration combined was under 3 billion current dollars in fiscal year 2008—an amount of money corresponding to about 2 cents per gallon on U.S. gasoline sales. (Throughout the remainder of this chapter, constant 2007 U.S. dollars are used unless stated otherwise.) The combination of state, local, and private sector spending on RD&D in the domains of energy supply and improving end-use efficiency is probably no more than twice the federal figure, hence under $6 billion per year.[1] A U.S. total of $9 billion per year or less for public and private ERD&D combined would correspond to less than 1 percent of the amount that the country was paying for energy at retail in the same period. For comparison, average R&D expenditures for all U.S. manufacturing sectors are around 4 percent of revenues, and some high-tech sectors such as pharmaceuticals, software, and computer chips reach the range of 10–15 percent of revenues. Another instructive comparison is that the $9 billion upper estimate for annual public and private ERD&D expenditures in the United States corresponds to only about six days' worth of U.S. oil imports at $130 per barrel.

1. Accurate estimates for private sector RD&D on energy are impossible to construct, not least because of the impossibility of determining, from publicly available data, what fraction of the large R&D expenditures of automobile companies and other manufacturers of energy-using goods can be attributed to efforts to improve the energy end-use efficiency of these products.

It might be supposed from such comparisons that increasing national investments in ERD&D should be an easy matter, but obviously it has not been. Fluctuations in ERD&D investments have accompanied oil price shocks and other causes of varying enthusiasm and optimism about improving the menu of energy-technology options. But the overall state of U.S. ERD&D in the early years of the twenty-first century remains what it has long been—woefully inadequate in relation to the challenges and opportunities the sector presents, despite a decades-long chorus of exhortation for more funding from the science and technology community, and a more or less steady increase in the evidence about the economic, environmental, and national security perils of continuing reliance on the currently available options and incremental modifications of them.

Many explanations for the evident difficulty of strengthening public and private efforts in ERD&D have been offered and analyzed. These include lack of sufficient private incentives and public and policymaker political will for investing in addressing the externalities and public goods liabilities of the conventional energy supply mainstays; the nature of energy as a commodity, thus subject to price fluctuations that impose large uncertainties on the returns to be expected from investment in innovation; the high capital cost and slow turnover time of typical energy facilities; the strong economies of scale in many energy supply technologies, which mean that demonstration projects large enough to establish economic competitiveness (or the lack of it) for a new technology are very costly; the "chicken and egg" problem associated with the large investments in new infrastructure needed to make some of the new technologies effective at scale (such as hydrogen pipelines and filling stations, pipelines for CO_2 sequestration, and additional transmission capacity for large-scale wind power); the financial pressures on corporations to allocate resources to functions with more predictable benefits than RD&D for the short-term bottom line; and the inherent difficulty of answering the question, with respect to investment in innovation, of "How much is enough?"

In the United States in particular, these generic obstacles to increasing ERD&D efforts are compounded by a widespread belief that the U.S. Department of Energy (DOE) is an especially cumbersome government bureaucracy that has too many nonenergy responsibilities and might not be able to spend increases in public funding for ERD&D in an effective manner, by the fact that federal ERD&D funding is in direct competition with politically more popular highway and water projects in the relevant congressional appropriations subcommittees, and by strident debates among purported experts about whether such problems as climate change and overdependence on oil are really problems at all.

Whatever the relative responsibility of these and other factors for the mismatch between current levels of effort in energy technology innovation (ETI) and the needs for improved technologies imposed by the very real energy problems that the United States and the world currently face, what is clear is that the mismatch is not fixing itself and must therefore be seen as a major challenge for policy. And what is needed from policy, of course, is not just more public resources and private incentives for ETI but improvement in the management and coordination of the efforts, including more—and more effective use of—partnerships (local-state-federal, university-industry-government, international, and combinations of these).

In the remainder of this chapter, we elaborate on the dimensions of this challenge and the available approaches for meeting it as follows:

—Section 2 expands on the nature of the main energy-related challenges faced by the United States, in the global context.

—Section 3 defines energy technology innovation, discusses its indispensable role in addressing the challenges, and expands on the reasons for the inadequacies in the ETI efforts of the U.S. public and private sectors.

—Section 4 addresses the uncertainties and time lags associated with reaping the benefits of ETI, and derives from this discussion some conclusions about the importance and urgency of decisive government action in ETI policy.

—Section 5 describes the technology-push and market-pull ETI policy options that could be used to shape and accelerate ETI in the United States relative to what is expected in the absence of such policies.

—Section 6 provides some overall conclusions.

Pressing Energy Challenges

Energy is critical in today's world through its effect on the economic, environmental, and sociopolitical dimensions of human well-being.

Economically, energy is an indispensable ingredient of basic material well-being and economic growth. Expenditures on it usually account for 7–10 percent of a country's GDP. It typically plays an even larger role in international trade and the associated balance of payments, and energy costs can powerfully affect economic competitiveness among regions and countries. In addition, energy equipment is a major, global, high-technology market.

Environmentally, the current technologies of energy supply are the dominant sources of many of the most dangerous and difficult environmental problems, from the very local (indoor air pollution from burning biomass and coal in inefficient stoves in badly ventilated residences) to the regional (outdoor air pollution, water pollution, and acid rain) to the global (climate change driven largely by greenhouse gases from the fossil fuel system).

Politically, control of energy resources is a source of political as well as economic leverage and power. Controlling or ensuring access to such resources has been a source of military conflict and could be again. Windfall energy revenues have been magnets in many places for corruption and have fueled arms buildups and paid for fomenting and carrying out terrorism. Nuclear energy, while avoiding many of the environmental problems of fossil fuels, has the potential for misuse as a source of nuclear weapons.

The challenges arising from these interactions are many and mostly obvious.

Economic Challenges

The fraction of GDP devoted to energy expenditures has been rising for the United States and most other countries, driven above all by the rising cost of oil but also by the rapid escalation in the cost of building electric power plants, which, in turn, has caused electricity prices to increase. When energy costs increase too much or too fast, of course, the result is inflation, recession, and sacrifice, above all for the poor.

Aside from the economic challenges related the high U.S. dependence on foreign oil (discussed in chapter 4), the other major economic challenge is that U.S. companies are at risk of arriving late to the global market for advanced energy technologies. In a world where energy needs could well be more than 50 percent higher in 2030 than they are today, the United States could lose substantial commercial opportunities if it does not manage to capture a significant share of the market in energy supply technologies (recently estimated at $440 billion per year in 2004 and growing).[2] Although the United States has some of the world's best research groups and laboratories, its industrial sector is currently not leading the world in the manufacture and deployment of most of the advanced energy technologies destined to claim a growing share of this market. For example:

—Eight of the top ten wind turbine manufacturers are European, and in 2005 they accounted for 72 percent of the global market, which was worth $23.2 billion in 2006.[3]

2. International Energy Agency (2007); National Commission on Energy Policy (NCEP 2004).
3. European Wind Energy Association, "Strategic Overview of the Wind Energy Sector" (www. ewea.org/index.php?id=195 [December 2008]).

—None of the top five global producers of photovoltaics (PV)—a sector with a market worth of $12.9 billion in 2007—were based in the United States in 2007.[4]

—Of the thirty-five new nuclear plants that were under construction as of July 2008, none were based in the United States, and the only generation III reactors in the world are located in Japan.[5]

—Currently two Japanese producers (Panasonic EV Energy and Sanyo) share over 85 percent of the world's market of batteries for hybrid electric vehicles, which use nickel metal hydride technology. This market was estimated to be worth almost $620 million in 2006 and is expected to grow up to $2.3 billion by 2013.[6]

A notable exception worth mentioning is biofuels. The United States surpassed Brazil in 2006 and, with 26 billion liters produced in 2007, is the world's largest producer of fuel ethanol. Biodiesel production in the United States is the second largest in the world, although it is still less than one-third of production in Germany.[7] More important, the United States is the world leader in the development of advanced biofuels technologies with large potential advantages over those in widespread use today.

An important reason for focusing on potential market share in advanced energy technologies, of course, is that this is not just a matter of balance of payments and economic growth per se, but also a matter of high-quality jobs for Americans.

Environmental Challenges

Global climate change is increasingly recognized as the most dangerous and probably the most intractable of all of the environmental impacts of human activity.[8] As discussed in chapter 2, there is no longer any reasonable doubt that the earth's climate is changing at a pace that is highly unusual against the backdrop of natural climatic variations.[9] There is also widespread scientific

4. Moran (2007); Renewable Energy Policy Network for the Twenty-First Century (REN21 2008). The top five global producers of PV cells were: Sharp (Japan), Q-Cells (Germany), Kyocera (Japan), Suntech (China), and Sanyo (Japan).

5. International Atomic Energy Agency, Power Reactor Information System, "Latest News Related to PRIS and the Status of Nuclear Power Plants" (www.iaea.org/programmes/a2/ [July 2008]); World Nuclear Association, "Reactor Database: Reactors under Construction" (www.world-nuclear.org/rd/rdsearch.asp [December 2008]). These generation III reactors are advanced boiling water reactors. They were built by a joint venture of GE, Hitachi, and Toshiba.

6. Anderman (2007).

7. REN21 (2008).

8. Holdren (2006).

9. Intergovernmental Panel on Climate Change (IPCC 2007a).

agreement that the accumulation of anthropogenic greenhouse gases (GHGs) resulting primarily from the combustion of fossil fuels is the main cause of these unusual climate changes, and that they are already having significant impacts on ecosystems and on human well-being.[10]

Aside from climate change, there are other serious environmental issues linked to energy resource extraction, processing, conversion, transport, and consumption, such as land degradation, water pollution, conventional air pollution, acid rain, depletion of aquifers and stream flow, and the impacts of the energy infrastructure on natural ecosystems.[11] But the climate change challenge is by far the most demanding environmental driver of the need for increased ETI.[12]

International Security Challenge

The economic and international security dimensions of the U.S. energy predicament are interconnected through U.S. economic vulnerability to oil supply disruptions and price shocks (see figure 5-1 and chapter 4).

The other major energy-related international security challenge is how to expand the use of low-carbon electricity from nuclear power—which, to have much effect on the climate problem, would need to happen not just in the United States but in many countries around the world—without accelerating the spread of nuclear weapons and increasing the risks associated with nuclear terrorism. The United States has 104 working power reactors with a capacity of 100 gigawatts of electricity (GWe), generating almost 20 percent of the country's electricity. The world has more than 400 power reactors totaling some 380 GWe of capacity and supplying about a sixth of global electricity.

But if nuclear energy is to generate as much as a third of the doubled world electricity demand that is likely by 2050 or sooner, over 1,500 GWe of nuclear capacity will be required. A considerable part of the increase would necessarily be in parts of the world that are of proliferation concern or that plan to reprocess spent nuclear fuel (and thereby create weapon-usable plutonium) or that are even more at risk from terrorists than the United States—or all of these.

---

10. Parmesan and Yohe (2003); IPCC (2007a); UN Environment Program, "Number of Flood Events by Continent and Decade since 1950," UNEP/GRID-Arendal Maps and Graphics Library (http://maps.grida.no/go/graphic/number-of-flood-events-by-continent-and-decade-since-1950 [December 2008]); UN Scientific Expert Group on Climate Change and Sustainable Development (2007). In 2004 CO_2 emitted from the use of fossil fuels made up 56.6 percent of total GHG emissions, measured in CO_2-equivalent (IPCC 2007b).
11. Hightower and Pierce (2008); UN Development Program (2000, chap. 3).
12. President's Committee of Advisors on Science and Technology (PCAST 1997, chap. 1). U.S. energy production accounted for 39 percent of freshwater withdrawals in the year 2000.

Figure 5-1. U.S. Primary Energy Consumption by Source and Sector, 2007

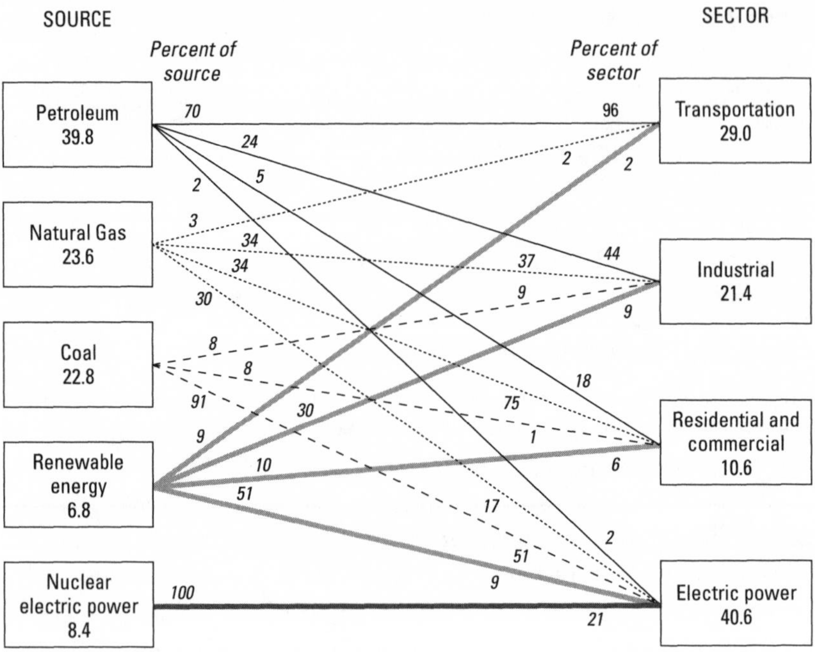

SOURCE   SECTOR

Percent of source   *Percent of sector*

Petroleum 39.8	70		96	Transportation 29.0
	24		2 2	
	5			
	2			
Natural Gas 23.6	3			
	34		37 44	Industrial 21.4
	34		9	
	30		9	
Coal 22.8	8		18	
	8		75	Residential and commercial 10.6
	91	30	1	
	9		6	
Renewable energy 6.8	10			
	51		17	
			2	
			51	
Nuclear electric power 8.4	100		9	Electric power 40.6
			21	

Source: U.S. Energy Information Agency, Office of Energy Statistics, "Energy Basics 101" (www.eia.doe.gov/basics/energybasics101.html [December 2008]).

If increases in nuclear energy of the indicated magnitude are to be achieved and sustained without disaster, more attention will need to be devoted not just to improving the technology of nuclear energy in ways that reduce its vulnerability to terrorist attack or misuse for nuclear weaponry but also to improving management and international oversight to these ends.[13]

Tensions among the Challenges

There are obvious interactions and often tensions among the economic, environmental, and security challenges associated with energy. One involves the potential security liabilities of expanding nuclear energy in order to help address the climate challenge, as just described. Another is the tension between ensuring the reliability and affordability of transport fuels and min-

13. Deutch and Moniz (2003).

imizing GHG emissions. The United States could try to increase the reliable supply of energy for transport by increasing domestic production of conventional oil and natural gas (usable in vehicles as compressed natural gas); encouraging increased production of unconventional oil (for example, from oil shales and tar sands); producing synthetic fuels from coal and biomass; increasing the production of biofuels; increasing electricity production from coal, nuclear, wind, or solar energy in order to charge plug-in hybrid or all-electric vehicles; and increasing the efficiency of liquid-fueled transport technologies in ways other than the use of hybrids. With the exception of the efficiency options and expanded use of nuclear, wind, and solar energy, however, with current technologies, all of the rest of these options would perpetuate CO_2 emissions from the transport sector and many would increase them unless equipped with CO_2 capture and sequestration technologies that are not yet available.

Need for Energy Technology Innovation

ETI is the set of processes leading to new or improved energy technologies that can augment energy resources, enhance the quality of energy services, and reduce the economic, environmental, or political costs associated with energy supply and use.[14] Like all types of technology innovation, ETI is characterized by research, development, demonstration, and deployment phases and the presence of multiple dynamic feedbacks between them.[15]

Energy Technology Innovation Is Essential to Address the Energy Challenges

ETI can help mitigate the economic liabilities of the current heavy dependence on oil and natural gas by improving energy efficiency in the transportation, electricity production, industrial, and building sectors; finding ways to avoid and replace the use of oil for transportation, for instance, through plug-in electric and fuel cell vehicles and the production of liquid fuels from coal or biomass; and creating new building blocks for the chemicals industry from other sources, such as coal or biomass.

ETI is also crucial to ensuring that the United States profits from the increasingly substantial business opportunities in the energy technology sector. In today's global economy, where the comparative advantage of high-cost

14. NCEP (2004).
15. Gallagher, Holdren, and Sagar (2006).

countries lies increasingly in knowledge-based innovation activities, the United States must ensure that it is the technological leader in the increasingly knowledge-driven energy sector.[16]

Substantially reducing global CO_2 emissions will be costly.[17] Achieving as much reduction in these emissions as quickly as is desirable, from the standpoint of decreasing the risk of catastrophic degrees of climate change, is likely to depend on keeping mitigation costs at the low end rather than the high end of the range of possibilities. The extent of success of expanded ETI efforts will be a key factor (if not *the* key factor) in determining whether this happens.[18]

Through the development of improved nuclear reactor concepts and nuclear waste disposal and reprocessing technologies, ETI can also help the United States maintain its leverage in global discussions of nuclear technology standardization, reduce the risks of terrorist attacks on nuclear power plants and terrorist acquisition of plutonium, and ease the demand for permanent storage sites for spent fuel. In addition, by increasing its efforts to find proliferation-resistant waste reprocessing technologies and long-term solutions to waste disposal, the United States could reduce the proliferation risks that would come from installation of less proliferation-resistant nuclear energy facilities in other countries.[19]

Because of the diversity of energy's roles in society, the multiplicity of energy-related challenges, the complexity of their interactions, and the fact that no energy technology currently known or imagined has the versatility and other characteristics needed to address all facets of the energy situation at once, ETI policy has the task of improving existing energy technologies and developing new ones across a wide range of applications and approaches. As is often said, "there is no silver bullet"—no single new or improved energy technology on which hopes for answers to the big energy questions can or should be pinned. ETI policies must be formulated to accelerate the development and deployment of a wide portfolio of technologies

16. Audretsch (2006).

17. MIT used its Emissions Predictions and Policy Analysis model to estimate the welfare costs of various cap-and-trade proposals aimed at reducing emissions in the United States, consistent with global stabilization scenarios corresponding to 450 ppm. Its modeling results indicated that GDP losses could be between 0.5–1.8 percent. See Paltsev and others (2007, p. 53).

18. Demand-side measures clearly are also essential to decrease energy consumption and thereby the cost of reducing GHG emissions. Demand-side measures induce changes in habits and lifestyles by encouraging people to use public transportation, buy energy-efficient appliances, turn their lights off, purchase fewer disposable items, and so forth.

19. Currently, France, Japan, the United Kingdom, and Russia are the only countries with spent fuel reprocessing facilities.

addressing diverse needs across multiple time scales—short term, medium term, and long term.

Why Has the U.S. ETI Effort Been Inadequate?

Given the ability of ETI to simultaneously and significantly address the pressing economic, environmental, and international security challenges faced by the United States, one may wonder why policymakers have yet to take ETI policy actions commensurate with the challenges. There are, of course, several reasons for the inadequacy of U.S. ETI policy today.

First, there is the widespread lack of understanding, among policymakers and public alike, of the nature of energy challenges and the role of ETI in addressing them. This translates into a lack of political will and corresponding shortfalls in the policy measures enacted to close the gap between private interests in ETI (which are considerable but do not capture the externality and public dimensions of energy) and the public's interest in it.

Structural characteristics of parts of the energy industry have also contributed to suboptimal private investment in ERD&D. For example, the critical mass of funding needed to generate advances in many classes of energy technology exceeds the capacity of small companies, which leads to private underinvestment in ETI in the energy sectors in which small companies predominate. In addition, the uncertainties and long timescales characterizing much of the ETI enterprise cause shareholder-focused firms to confine R&D investments to projects offering quick returns with low risk; this syndrome afflicts large firms as well as small ones.

Another reason for the inadequacy of U.S. ETI policy is that energy is a commodity, and as such it is subject to price fluctuations and uncertainties on the returns to be expected from investments in innovation. During periods of low energy prices, the private sector has little incentive to invest in ETI. And although recent periods of high energy prices have created market space for investments in innovation in areas such as coal-to-liquids and biofuels, concern that prices will fall in the future still limits what is invested.

Furthermore, traditional, centralized energy supply facilities are characterized by high capital costs, strong economies of scale, and long turnover times, all of which inhibit the commercialization and penetration of alternatives.

Demonstration projects aimed at scaling up, testing, improving, and demonstrating the commercial viability of new technologies are essential in bridging the gap between R&D and commercial success, but they are costlier and "lumpier" than R&D and correspondingly more problematic for firms to fund. To deploy advanced technologies (for example, electric vehicles, fuel cell vehicles, large-scale wind or solar electricity, and carbon storage) on a significant scale, it is often necessary to have an underlying infrastructure— such as additional transmission and distribution lines, hydrogen pipelines and fueling stations, and CO_2 pipelines—or other advanced technologies, such as those needed for large-scale energy storage. But building the required infrastructure and investing in ancillary advanced technologies are expensive operations that are difficult to finance without some certainty about the size of the market that will be using the infrastructure.

The belief that the market is more efficient at commercializing energy technologies has been another factor in the decline of public support for ERD&D in the United States over the past three decades.[20]

Lack of private-sector certainty about what future public policies governing energy options will be continues to affect the private sector's willingness to invest in developing, demonstrating, and deploying advanced energy technologies.

Finally, an effective institution able to advocate for and implement a sounder ETI strategy has been lacking. While the DOE might be thought to have this capability, it has been hobbled both by not having all of the federal ETI portfolio under its jurisdiction (other parts reside in the Environmental Protection Agency and the Departments of the Interior, Commerce, and Agriculture, among others) and, even more important, by having too many non-energy missions claiming the attention of its management (for instance, nuclear weapons, environmental remediation, and basic science). This institutional failure, together with the short-term nature of the federal budget process, has made it difficult to create and sustain a coherent ETI policy and the budgets to support it. The intrinsic difficulty of determining how much money should be devoted to ETI and deep-seated differences of opinion about how such money should be allocated among the different energy technology sectors further complicate matters and, indeed, promote policy paralysis.

Although the aforementioned "causes" of inadequacy in the ETI effort— which for the most part apply not only to the United States but also to most of the other industrialized countries where the bulk of ETI effort has been

---

20. The advocates of the "let the market do it" argument were reinforced by the failure of two large-scale, DOE-funded energy-technology-demonstration projects at the end of the 1970s: the Clinch River Breeder Reactor and the Synthetic Fuels Corporation. The failure of the former project was due to overly optimistic engineering estimates of cost and technological readiness as well as overenthusiastic projections of need; that of the latter was due to falling oil prices and the fact that the project results were so dependent on DOE business practices that they were not credible to the private sector. See, for example, U.S. General Accounting Office (1996, p. 19); Ogden, Podesta, and Deutch (2008).

concentrated—have been largely understood and acknowledged for many years, little has been done to change the situation in any fundamental way.

Acting in Time to Strengthen the U.S. Energy Innovation System Is Crucial

The United States has reached a point where postponing action—either out of desire to know more about climate change or out of faith that new technologies will materialize automatically in time to reduce the costs of dealing with the U.S. energy challenges—is no longer an option. A concerted ETI effort must be set in motion over the next few years if the United States is to maintain its role as a flourishing and competitive economy, give the world a chance to prevent a climate disruption crisis of unmanageable consequences, and minimize the chances of international conflicts related to fossil fuels or nuclear energy. Long delay in starting the needed ETI effort has shortened the remaining window of opportunity.

Failure to take immediate action to strengthen the U.S. ETI system is likely to have a number of economic consequences. Growing global demand for energy and for low-carbon technologies is creating a large market for suppliers of advanced energy technologies. The International Energy Agency estimates that over $21 trillion will have to be invested in energy technologies between 2005 and 2040 to meet the world's energy demand in 2030.[21] Investments in clean energy are already growing at a fast pace worldwide: global public and private sector expenditures in clean energy increased by 60 percent from 2006 to 2007, reaching $148 billion.[22] Yearly investment in renewable energy, moreover, is expected to triple by 2012 to reach $450 billion and quadruple by 2020 to reach $600 billion.

Along with this growth in market opportunities, international competition to seize them is also growing. Of the total $118 billion spent on sustainable energy through venture capital (VC) deals, private equity, public markets, and asset finance worldwide in 2007, 47 percent went to Europe, 23 percent to the United States, and 22 percent to developing countries; China, India, and Brazil alone received 17 percent.[23] The share of total

21. International Energy Agency (2007).
22. Boyle and others (2008).
23. The source of this information is a study by New Energy Finance (Boyle and others 2008) commissioned by the UN Environment Program. The study defines sustainable energy as all biomass, geothermal, and wind generation projects of more than 1 MW; all hydro projects between 0.5 and 50 MW; all solar projects of more than 0.3 MW; all marine energy projects; all biofuels projects with

investments captured by emerging economies is growing very rapidly.[24] Evidently, other industrialized economies and emerging economies are determined to capture large shares of the growing energy technology market and are well on their way. Some specific examples:

—In 2005 Japan released its "Strategic Technology Roadmap in the Energy Field," identifying its short-, medium-, and long-term energy R&D needs, and in January 2008 it announced that it will spend $30 billion over the next five years in ERD&D and $10 billion assisting developing countries in the deployment of low-carbon technologies.[25]

—The European Union (EU) has embarked on a large effort to put ETI at the forefront of its policy agenda. Its Strategic Energy Technology Plan (SET Plan), which was released in November 2007, aims to support the EU's ambitious renewable energy goals and complement its Emissions Trading Scheme system.[26] The SET Plan's components include creation of a Steering Group on Strategic Energy Technologies, a European Energy Technology Information System, an increase in EU-wide funding for ETI, and the creation of the European Industrial Initiatives.[27] The SET Plan is still in a consensus-building state. In addition, recent developments indicate that the EU will provide very significant amounts of financial support for technology demonstration projects, such as carbon capture and storage, using the new entrants reserve of the EU Emissions Trading Scheme.[28]

---

a capacity of 1 million liters or more per year; and all energy efficiency projects that involve financial investors. The $118 billion figure excludes governmental and corporate RD&D and small-scale projects (approximately $36 billion) and reinvestment ($5.3 billion). Although the United States and Europe each received about half of the total $9.8 billion of investment in early-stage financing (that is, VC and private equity), Europe is leading the world in public market and asset finance investments, which can be considered a proxy for deployment.

24. Ibid.

25. See Japanese Ministry of Economy Trade and Industry, "Strategic Technology Roadmap (Energy Sector)—Energy Technology Vision 2100" (www.iae.or.jp/2100/main.pdf). See also Prime Minister of Japan and His Cabinet, "In Pursuit of 'Japan as a Low-Carbon Society,'" speech by H.E. Mr. Yasuo Fukuda, Prime Minister, at Japan Press Club (www.kantei.go.jp/foreign/hukudaspeech/2008/06/09speech_e.html [June 2008]).

26. In 2007 the EU committed to achieving the following goals by 2020: reducing its GHG emissions by 20 percent, increasing energy efficiency by 20 percent, increasing the share of renewables in overall EU energy consumption from 8.5 percent to 20 percent, and increasing the biofuel component of vehicle fuels up to 10 percent.

27. Commission of the European Communities (2007).

28. See question 7 in Europa, "Questions and Answers on the Revised EU Emissions Trading System," December 17, 2008 (http://europa.eu/rapid/pressReleasesAction.do?reference=MEMO/08/796).

The interaction between global markets and policies is far too complex to predict the exact fraction of the market—of the estimated $600 billion per year—that U.S. firms would be able to capture if the U.S. government changed course and made ETI policy a priority, but it seems reasonable to assume that the difference between action and inaction will be at least in the range of tens of billions of dollars.

In the same vein, failure to implement a comprehensive set of federal ETI policies would be a missed opportunity to create high-quality jobs in the United States. For example, a recent study estimated that if the U.S. government spent $100 billion in energy efficiency and renewable energy programs, 2 million jobs could be created.[29] Adopting a 20-percent renewable portfolio standard (RPS) could create 100,000 to 200,000 more jobs than using a 50–50 mixture of natural gas and coal power plants to meet that 20 percent of electricity demand.[30] The longer it takes to deploy renewable electricity technologies, the more conventional power plants will have to be built to meet demand and the longer the full benefit of new renewable energy jobs will be deferred.

For reasons discussed in detail in chapter 1, accelerating the development and deployment of low-carbon energy supply and end-use technologies is also urgent from a climate change perspective. And it is urgent, too, from the standpoint of addressing energy-related international security risks, even if these are harder to quantify than those associated with economic and environmental challenges.

Global demand for oil and natural gas is projected to grow faster than production in the decades immediately ahead, which can hardly fail to increase international tensions. The specific vulnerabilities and foreign policy liabilities for the United States due to its heavy dependence on oil and natural gas are on track to grow unless the country changes course. And the vulnerability of U.S. energy infrastructure to terrorist attack cries out for remedy.

Nuclear energy has the potential to contribute materially to the ability of the United States and other countries to meet expanding electricity needs while reducing GHG emissions. To make a really significant contribution to this challenge, however, nuclear energy would need to increase its share of

29. Bracken Hendricks, "The Green Road to Economic Recovery," testimony before the House Select Committee on Energy Independence and Global Warming, September 18, 2008 (www. americanprogressaction.org/issues/2008/pdf/hendricks_testimony.pdf).

30. Kammen, Kappadia, and Fripp (2004). These jobs include fuel procurement, operation and maintenance, and manufacture and installation.

global electricity production from one-sixth in 2000 to a third or more in 2050 and 2100. Getting to a third, under business-as-usual electricity growth, would involve expanding nuclear energy from 350 GWe in 2000 to 1,700 GWe in 2050 and 3,500 GWe in 2100, with most of this capacity increase occurring in industrializing countries.[31] An expansion of this magnitude without an acceleration of improvements in nuclear energy technology, more rigorous regulation, and more creative approaches to avoiding an accompanying aggravation of proliferation dangers would likely result in increased risk of major accidents, proliferation, and nuclear terrorism, probably accompanied by a withdrawal of public consent for expanded reliance on this option.[32]

Policy Tools for Energy Technology Innovation and Recommendations

Energy technology innovation policies are those government actions aimed at shaping the direction and pace of such innovation. Although it is a simplification, given the nonlinear nature of innovation, these policies can be divided into *technology-push* and *market-pull* ETI polices (see figure 5-2), each essential if ETI is to reach its potential for helping with the energy-related economic, environmental, and national security challenges the country faces.

Federal policies on the technology-push side can be divided into those that fund ERD&D activities and directly encourage an increased participation of the private sector and an increased cooperation between the United States and other countries, and those that increase the quality and the quantity of the ETI workforce. The policies available on the market-pull side can be divided into those that encourage the early deployment and widespread diffusion of new energy technologies through direct expenditures, tax-related expenditures, or financial support; those that spur ETI by setting technology performance standards; and those that impose climate regulations to encourage ETI and steer it in a direction that correctly accounts for the negative environmental externalities of GHG emissions. As hinted earlier, even this taxonomy of ETI policies is not exhaustive, but it serves to illustrate the richness of avenues for influencing the pace of change in a country's energy technology base.

31. Deutch and Moniz (2003).
32. Bunn (2008).

Figure 5-2. Types of Technology-Push and Market-Pull Policies Shaping ETI

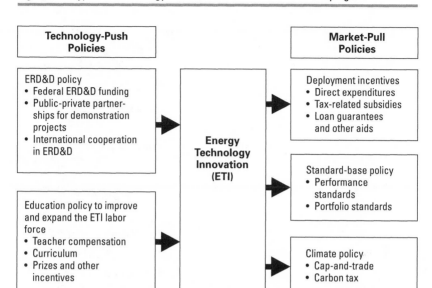

Energy Research, Development, and Demonstration Policy

Any effective portfolio of policies for addressing ETI must take account of needs and opportunities in both the public and private sector, the interactions between them, and the role of international cooperation in ETI. In what follows we treat these matters in turn.

Public Sector. Since its creation by President Carter in 1977, the Department of Energy has been the main source of federal funding for basic energy sciences (BES) and applied energy technology research, development, and demonstration. For example, in fiscal year 2007, DOE provided $2.5 billion for ERD&D, while the combined ERD&D funding from the U.S. Department of Agriculture, U.S. Geological Survey, and Defense Advanced Research Projects Agency amounted to $76 million, or 3 percent of the total federal ERD&D expenditures.[33] Funding for BES is completely centralized in DOE and amounted to $1.1 billion in fiscal year 2007.

Given the prevalence of DOE in supporting ERD&D in the United States, the decline in DOE ERD&D funding observed over the past thirty years is particularly worrying (figure 5-3). The DOE's ERD&D funding reached a

33. Gallagher (2008); U.S. Energy Information Agency (EIA 2008).

Figure 5-3. DOE ERD&D Spending, Fiscal Year 1978 to Fiscal Year 2009 Request[a]

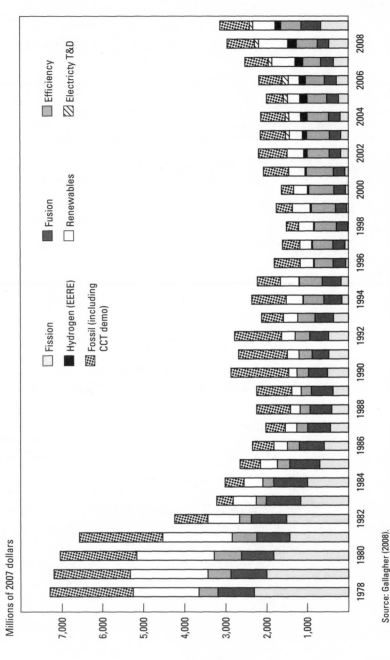

Source: Gallagher (2008).
a. CCT, clean coal technology; EERE, DOE Office of Energy Efficiency and Renewable Energy; T&D, transmission and distribution.

maximum of $7.3 billion in 1978 and a low of $1.6 billion in 1998, and it increased gradually to $2.5 billion in 2007, which is still below the 1978 level by almost a factor of 3. During this time the U.S. GDP grew by a factor of 2.3, from $6 to $14 trillion; thus, as a percentage of GDP, DOE ERD&D funding decreased very sharply, from 0.12 percent to 0.02 percent.[34]

The decline in public sector funding for ERD&D over the past thirty years is not unique to the United States; it is shared by all International Energy Agency member countries (essentially the Organization for Economic Cooperation and Development [OECD]) with the exception of Japan.[35] It is notable that Japan now spends more public funds on applied energy technology RD&D than the United States does, despite Japan's more than twofold smaller economy. (See figure 5-4.)[36]

The quantity and allocation of public ERD&D funding are not the only important factors to consider when assessing a portfolio of ETI investments. Measures of the productivity of those investments are also important. We are aware of just two recent major studies that tried to assess the productivity of public investments in ETI in the United States. The first of these analyzed the economic benefits of twenty of DOE's most successful energy efficiency and renewable energy programs between 1990 and 1998.[37] The twenty programs were estimated to have resulted in energy savings of 5,500 trillion BTUs— equivalent to almost 14 percent of primary electricity consumption in the United States in 2007—over the lifecycle of the deployed advanced technologies in the 1990s.[38] By using an average value of the cost to consumers of a BTU of energy in 1998, DOE estimated that the economic savings that resulted from those twenty programs were approximately $37 billion. The value of those economic savings was more than sufficient to compensate for the $23.7 billion spent by DOE between 1978 and 2000 in all of its energy efficiency and renewable energy programs.

The second study, by the National Research Council (NRC), evaluated DOE's fossil energy and energy efficiency programs between 1978 and 2000.

34. U.S. Department of Commerce, Bureau of Economic Analysis, "National Income and Product Accounts Table" (www.bea.gov/national/nipaweb/TableView.asp?SelectedTable=5&FirstYear= 2007&LastYear=2008&Freq=Qtr).

35. Runci (2005).

36. To correctly interpret the data in figure 5-4, one must take into account the fact that in the case of the United States, the "other research" category, which accounted for 42 percent of DOE's expenditures in ETI in 2006, includes U.S. expenditures in BES, biological research, and environmental research.

37. DOE (2000).

38. EIA, Office of Energy Statistics, "Energy Basics 101" (www.eia.doe.gov/basics/energy basics101.html [December 2008]).

Figure 5-4. Reported 2005 Public Expenditures in ERD&D and "Other Research" by Twelve of the Twenty-Five EU Member States (MS), Japan, and the United States, and Funding for Energy R&D by the EU's Sixth Framework Program for Research (FP6)

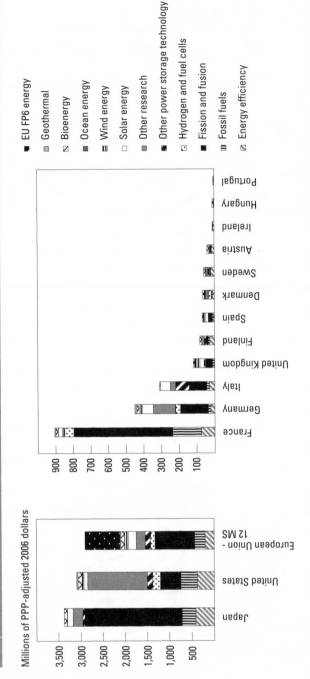

Sources: IEA, "R&D Statistics, Access Database" (www.iea.org/textbase/stats/rd.asp); Commission of the European Communities (2007, commission staff working document, figure 19), available at http://ec.europa.eu/ energy/technology/set_plan/doc/2007_capacity_map.pdf.

The NRC found that during this period, six energy efficiency programs—which had been cofounded by the DOE and the private sector at a combined cost of $0.5 billion—had an economic benefit worth $36 billion.[39] This benefit significantly exceeded the $9.5 billion spent by DOE on all energy efficiency programs between 1978 and 2000.[40] The NRC study also found that the combination of federal funding for ERD&D, financial incentives for adoption of new technologies, and efficiency standards for buildings and equipment had been a major driver of the success of the efficiency programs.[41] The refrigerator (see figure 5-5), electronic ballast, and low-emissivity window glass programs are examples of technologies that are widely used today thanks to a *virtuous cycle* by which RD&D successes allow the introduction of more stringent performance standards, and vice versa, higher standards encourage more RD&D successes. These case studies highlight the importance of designing comprehensive policy packages.

Virtually every group that has looked recently at the record of U.S. public investment in ERD&D in relation to needs and opportunities has ended up calling for an expanded, better managed, and better evaluated effort. In its monumental 1997 study of federal energy R&D strategy, the President's Committee of Advisors on Science and Technology (PCAST) called for nearly a doubling of DOE ERD&D spending over a five-year period, along with

39. NRC (2001). There were six programs selected for detailed cost-benefit analysis: advanced refrigerator-freezer compressors, electronic ballast for fluorescent lamps, low-emissivity glass, advanced lost foam casting, oxygen-fueled gas furnaces, and advanced turbine systems. They included programs in the building technology, vehicle technology, and industry subprograms. The economic benefits were the energy cost savings between 1978 and 2000 (net of R&D costs, extra capital, and labor costs) compared with the next best alternative and calculated on the basis of the life-cycle of investments. The economic benefits were calculated assuming that without the DOE investment, the technical innovations would have followed the same market penetration displaced by five years.

40. Ibid. The economic return of $13.2 billion to the eight fossil energy programs assessed, which had cost a total of $4.2 billion to DOE and private sector partners, were modest in comparison, and they did not allow the committee to claim that those eight programs had produced enough returns to cover the costs of the $18 billion invested by DOE over twenty-two years in fossil energy programs. These economic returns, however, did not include environmental and energy security benefits, which the NRC also estimated, or the knowledge benefits arising from the fact that some of the technologies developed within these DOE fossil energy programs might be of use in the future. Some examples of technology areas with knowledge benefits that might become useful in today's world of high oil prices are integrated gasification combined cycle, oil shale extraction, and indirect coal liquefaction. Finally, the NRC study almost certainly underestimated even the short-term economic benefits through its assumption that innovations made with government funds would have occurred in the private sector without cost to the government five years later.

41. Ibid.

Figure 5-5. Evolution of Annual Electricity Use, Cost, and Volume from Household Refrigerator-Freezers, 2003[a]

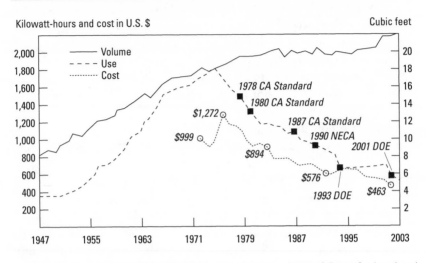

Source: Geller and Goldstein (1999); NCEP (2004, Technical Appendix—Chapter 6); Brown, Southworth, and Stovall (2005, p. 57, figure 15).

a. Average energy use per unit per year, cost in 2003 dollars, and adjusted average volume. CA = California; NECA = National Electricity Contractors Association.

more coordinated management and improved evaluation.[42] Schock and others estimated that the amount of public ER&D funding warranted from an "insurance" standpoint in light of the risks from climate change, oil price shocks, urban air pollution, and other energy disruptions was $6 billion to $9 billion, in the range of four to six times what was being spent at the time.[43] The 2004 report of the independent, bipartisan National Commission on Energy Policy recommended a doubling of ERD&D expenditures over the period 2005–2010, a recommendation reiterated in the commission's 2007 update.[44] Nemet and Kammen compared the need for an expansion in public ERD&D with that of the Apollo and Manhattan projects and called for an increase in ERD&D funding by a factor of 5 to 10.[45] Ogden, Podesta, and Deutch recently called for at least a doubling of DOE's ERD&D budget.[46]

42. PCAST (1997, chap. 1).
43. Schock and others (1999).
44. NCEP (2004, 2007).
45. Nemet and Kammen (2007).
46. Ogden, Podesta, and Deutch (2008).

Private Sector. It has been estimated that about two-thirds of U.S. R&D funding for all purposes comes from the private sector.[47] Some recent estimates for the case of energy, however, have suggested that the private sector only contributes between one-third and one-half of total U.S. ER&D expenditures and that private sector investments in this domain have fallen over the past two decades by something like a factor of 4.[48] The decrease in private sector ER&D funding recorded by Nemet and Kammen does not include funding from VC firms. With a value of over $2.7 billion, VC firm investments in the clean energy sector in 2007 were substantial, but the largest part of those funds probably did not go to ER&D.[49]

Attempts to track private sector ERD&D expenditures are notoriously difficult. They are plagued by lack of access to complete data as well as by ambiguities about what proportion of research on technologies that use energy—such as vehicles—should be classified as "energy research."[50] Many in the energy industry argue that publicly available estimates of the private sector's ERD&D are far too low and that assertions of a large real decline in recent decades are exaggerated. We take the range of possibilities for the private sector's contribution to U.S. ERD&D expenditures to be, in round numbers, between one-third and two-thirds of the total.

Although this is a large uncertainty, even at the high end the ERD&D investments of the private sector can plausibly be argued as being inadequate, not only because they are low among virtually all high-tech industries as a proportion of revenues but, more tellingly, because the recent and current pace of advance is so clearly short of what the challenges require. Boosts in the incentives for private sector RD&D in this domain are quite evidently warranted.

Public-Private Interaction. One might be tempted to think that having both the private and the public sectors ramp up their ERD&D efforts simultaneously would put them in conflict with one another—that public ERD&D would "crowd out" private investment—but recent analysis suggests that this is not the case. Public funding for R&D in the United States and in particular for energy R&D does not appear to be substituting for or crowding out private sector investment.[51] Of course, government can take steps explicitly designed to encourage increased private sector activity in ETI, and most

---

47. American Association for the Advancement of Science, "R&D Budget and Policy Program" (www.aaas.org/spp/rd/guitotal.htm [January 2009]).
48. Nemet and Kammen (2007); Dooley (1999).
49. Makower, Pernick, and Wilder (2008).
50. Gallagher, Holdren, and Sagar (2006).
51. David, Hall, and Toole (1999); Nemet and Kammen (2007).

recent studies of the matter have recommended that, along with ramping up direct public funding of R&D (some of which gets performed by the private sector), the federal government should provide increased tax incentives for the private sector to increase the ERD&D that it undertakes.

Of particular importance and sensitivity is the question of the respective roles of the public and private sectors in the demonstration and early deployment phases of ETI. The poor record of governments in "picking winners" in the energy technology domain has tended to support the view that, in light of the private sector's necessarily closer familiarity with market forces and conditions, the role of government should get smaller and that of the private sector larger in the demonstration and early deployment phases of the innovation process.[52] On the other hand, as discussed in some detail in the 1997 and 1999 PCAST reports, the relevance of the externality and public goods rationales for accelerating ETI extends beyond the research and development phases: if these rationales justify government investment in R&D, they also justify some government engagement in helping to ensure that the products of R&D most germane to these rationales get translated into practice.

Indeed, as discussed in the PCAST reports and elsewhere, government involvement in the demonstration and early deployment stages of technology innovation can be critical in overcoming some of the barriers between R&D and commercialization, which together have been termed "the valley of death."[53] Increasingly, the creative use of public-private partnerships has been seen as the solution to the dilemmas attendant on the roles of the public and private sectors as innovation moves into the demonstration and deployment phases.

International Cooperation. The importance of international cooperation in ERD&D arises from the various inherently global problems and opportunities of energy systems—climate change, nuclear proliferation, effects of global energy resource and energy technology markets, and the widespread economic and security benefits of providing access to modern energy for the poorest third of humanity—as well as from the obvious benefits of sharing the costs and risks of the innovation process.[54] Cooperation on fundamental research with leading scientific groups from other countries does not involve intellectual property rights conflicts and has the potential to accelerate technical progress. Cooperation with other countries can accelerate innovation in those countries that would have value for the United States

52. Brooks (1967).
53. See, for example, Gallagher, Holdren, and Sagar (2006).
54. PCAST (1999).

and the rest of the world (for example, CO_2 capture and sequestration for coal-burning power plants in China and India, and adoption of proliferation-resistant nuclear energy technologies everywhere). Cooperation on ETI with other countries also provides knowledge about and access to wider markets for U.S. innovations.

ERD&D Recommendations

The decline in public and private ERD&D funding, increased international competition, observed returns to government ERD&D programs, and the growing security, economic, and environmental challenges ahead all underscore the need for a bold and coordinated U.S. ERD&D policy. It should include the following elements:

—Increasing DOE's ERD&D budget by at least a factor of 2 in real terms over the next few years, pushing these expenditures above $6 billion per year. This increased budget should be made more reliable to allow for better project planning and continuity.

—Directing a large fraction of the budget increase to public-private partnerships for technology demonstration projects, to technologies capable of dealing simultaneously with the largest challenges, and, of course, particularly to high-risk, high-payoff technologies not receiving adequate attention from the private sector.

—Adopting a portfolio approach to investing in ERD&D projects, taking account of which technologies are likely to be substitutes or complements, stages of technology development, the criticality of federal RD&D investments, and measures of the potential benefits of each technology.

—Increasing funding for the BES program, focusing on cross-cutting (integrative) research and novel concepts.[55]

—Strengthening the organizational structure and management procedures of DOE to effectively manage this expanded and more sophisticated federal ERD&D enterprise. Particular areas of improvement for DOE should include increased communication between BES and applied ERD&D programs; better coordination between the activities of DOE and other federal agencies (in particular the Departments of Commerce, the Interior, and Agriculture, as well as the National Science Foundation, Environmental Protection Agency, and U.S. Agency for International Development); and a more

---

55. The Energy Frontier Research Centers constitute an excellent recent initiative of the BES program at DOE. They are selected by scientific peer review and funded at $2–5 million per year over a 5-year period. They will be awarded to groups conducting fundamental research focusing on the "grand challenges" identified by a major planning effort undertaken by the scientific community.

systematic effort to improve and increase the number of relationships with industrial partners.[56]

—Considering creation of an independent, complementary entity for promoting and funding demonstration and early commercialization efforts, along the lines of the Energy Technology Corporation proposed by Ogden, Podesta, and Deutch.[57]

—Making private sector R&D tax credits permanent.

—Increasing funding for international cooperation in fundamental research and applied ERD&D by at least a factor of 3 and adopting a coherent international cooperation strategy. This funding should be directed largely to fundamental research with countries that are advanced technologically, such as Japan and the European Union, and to more applied research and demonstration projects in developing countries with rapidly expanding energy markets.[58]

Education Policy

Without educated people carrying out the ETI enterprise and making rational decisions regarding the acquisition and use of energy technologies, funds allocated to energy RD&D will not yield the hoped-for benefits. In other words, the "human input" is essential to meet current and future energy challenges.

A number of indicators suggest that the United States is not doing enough to create the human resources the country needs to be a competitive innovator in the energy technology field. The 2005 report *Rising above the Gathering Storm*, from the National Academy of Sciences, found that although the United States still excels in higher education and training, and although U.S. scientists and engineers still lead the world in publishing results, other countries are "catching up."[59] Forty percent of the increase in global publishing in science and engineering between 1988 and 2001 came from Japanese, western European, and Asian emerging economies; U.S. publications have remained essentially constant since 1992. In addition, in 2001

56. Gallagher, Frosch, and Holdren (2004); NCEP (2004); NRC (2001).

57. Ogden, Podesta, and Deutch (2008). An Advanced Projects Research Agency for Energy (ARPA-E), modeled after DARPA, was authorized in 2007. The extent to which this entity might fill the needed role remains unclear at the time of this writing.

58. An example of the type of projects that should be pursued with countries that are technologically advanced is the International Thermonuclear Experimental Reactor. This project's partners—the United States, Japan, the EU, China, India, and South Korea—are sharing the cost of developing fusion energy technologies that could (in the very long term) radically change the world's energy system.

59. National Academy of Sciences (2005).

the U.S. trade balance for high-technology products became negative, and it was still negative in 2006.[60]

The increased international competition in the fields of science and engineering poses challenges to the U.S. education system. Because of the comparative disadvantage of U.S. K–12 education (as highlighted by results from the Program for International Student Assessment [PISA]), there is limited graduate interest in science and engineering (S&E) majors and significant attrition among S&E undergraduate and graduate students; in some instances S&E education is inadequately preparing students to work outside universities.[61] Most important, the National Academy of Sciences report identified the importance of optimizing the country's science and engineering human resources to respond to the nation's need for clean, reliable, and affordable energy and to create high-quality jobs for Americans.

Education Recommendations

Many of the education-related recommendations of the National Academy of Sciences report, which are summarized below, are essential to accelerate ETI and ensure the sufficient size and continuity of the ETI enterprise.

—Improve K–12 science and mathematics education by recruiting more science and mathematics teachers; strengthening the skills of science and mathematics teachers by supporting continuing education in the form of summer institutes, master's programs, and other types of training; and creating a voluntary national world-class science and math curriculum.

—Encourage high school students to pursue S&E undergraduate degrees by awarding four-year scholarships for students interested in the physical and life sciences, engineering, and mathematics.

—Encourage undergraduate and graduate students to pursue careers in S&E research by increasing the number of graduate fellowships for U.S. citizens in "areas of national need," increasing the number of grants for outstanding early-career S&E scholars, and creating other incentives, such as prizes.

---

60. National Science Foundation (2008, chap. 6, figure 6-23).

61. Baldi and others (2007). The 2006 PISA data collection effort found that American fifteen-year-olds had average science and mathematics literacy scores below the average of all OECD countries and below some non-OECD countries. PISA is sponsored by the OECD and assesses the reading, science, and mathematics literacy of fifteen-year-olds in fifty-seven jurisdictions, thirty of which are OECD jurisdictions and twenty-seven of which are not. Unlike other tests, the PISA assessments measure application of knowledge, rather than curricular outcomes, and are thus useful to make comparisons across countries. In 2002 science, technology, engineering, and mathematics degrees accounted for 17 percent of all first university degrees awarded, compared to an international average of 26 percent. Kuenzi (2008).

—Ensure that the United States remains the most attractive place in which to study and perform research by, among other things, improving the visa process for international students and scholars.[62]

Many of the aforementioned recommendations were authorized in 2007 by the America COMPETES Act; the 10,000 Teachers, 10 Million Minds Act; and the Sewing the Seeds through Science and Engineering Act. It is now essential to ensure that these initiatives are sufficiently funded, progress is monitored, and feedback is being used effectively to improve the programs.

Deployment Incentives

The energy challenges faced by the United States and the world will not be adequately addressed unless new energy technologies are deployed at a large scale. It follows that market-pull policies are necessary complements to the technology-push policies discussed until now, and that there should be coordination between both efforts to avoid wasting resources.

Once the technical feasibility of an energy technology has been demonstrated at scale, other barriers often still prevent the technology from competing in the marketplace and achieving the necessary large-scale deployment. These barriers include cost-effectiveness; intellectual property issues; fiscal, regulatory, and statutory barriers; lack of information; and policy uncertainty.[63] Deployment incentives are aimed at overcoming these barriers, and their nature is as varied as the barriers they are aiming to overcome. It is, therefore, very difficult to obtain an accurate picture of energy technology deployment policies in the United States, especially when one takes into account the fact that states also have their own deployment incentives. Nonetheless, it is useful to review federal incentives aimed at dealing with the most pervasive cost-effectiveness barriers because effectiveness considerations require that the federal government ultimately be in charge of adopting a coordinated energy technology strategy in basic research, ERD&D, and deployment.[64]

The U.S. Energy Information Administration categorized federal subsidies to energy markets into RD&D expenditures and deployment expenditures, the latter divisible into direct expenditures, tax-related subsidies, and financial support.[65] In 2007 the largest subsidies for electricity production were being directed to electricity from refined coal ($29.8 per MWh), fol-

62. National Academy of Sciences (2005).
63. Brown and others (2007).
64. According to an analysis from the Oak Ridge National Laboratory. Ibid.
65. EIA (2008).

lowed by wind ($24.3 per MWh), solar ($23.4 per MWh), nuclear ($1.6 per MWh), and landfill gas ($1.4 per MWh).[66] The largest energy subsidies per million BTUs unrelated to electricity production were being directed to bio-fuels ($5.7 per million BTUs), solar energy ($2.8 per million BTUs), and refined coal ($1.3 per million BTUs). A breakdown of the subsidies by category and energy source gives useful insights regarding the role of large deployment subsidies in getting advanced energy technologies off the shelf (figure 5-6).[67] Tax-related subsidies dominate in terms of size of the expenditures, but, most important, the largest deployment subsidies are not targeted enough to adequately buy down the costs of advanced technologies or help them reach the tipping point into widespread diffusion. Specifically:

—The largest subsidy ($3.0 billion) was the Volumetric Ethanol Excise Tax Credit (VEETC). VEETC has been identified as highly ineffective because the cost of this subsidy in 2006 exceeded $1,700 per ton of CO_2 avoided and $85 per barrel of oil replaced.[68]

—The second-largest subsidy was the $2.4 billion spent in refined-coal electricity from the Alternative Fuel Production Tax Credit. This tax credit expires at the end of 2009 and was not targeted to promote the deployment of advanced coal technologies.

—With a value of $0.9 billion and $0.8 billion, the "expensing of exploration and development costs" and the "excess of percentage over cost depletion" tax expenditures were the third- and fourth-largest expenditures, respectively. Both of them fall into the oil and gas category and are more effective at encouraging an increase in domestic oil production than stimulating the deployment of advanced technologies.

66. The Internal Revenue Service defines synthetic fuels as coal that has undergone a refining process that has produced a "significant chemical change"—a condition that is satisfied by spraying coal with limestone, acid, diesel, or other substances. Until December 31, 2007, synfuels qualified as "refined coal" provided the facility was placed in service before June 30, 1998. The definition of refined coal was revised in the American Jobs Creation Act of 2004. The new definition is that it is a gaseous, liquid, or solid synthetic fuel derived from coal sold by the taxpayer with the expectation that it will be used to produce steam, certified as resulting in a qualified emission reduction, and produced in such a manner as to result in a product with a market value 50 percent greater than that of coal, excluding increases caused by the materials added during its production.

67. Note that the breakdown excludes $1.9 billion in subsidies to energy efficiency programs in the industrial, transportation, and residential sectors, and $1.0 billion to improve transmission and distribution.

68. Metcalf (2008). This criticism does not imply that the VEETC has had no effect in bringing about private sector investments in biofuels but rather that it would be more effective if it were more targeted.

Figure 5-6. Direct Funding for ERD&D from DOE and Other Agencies, and Deployment Incentives for Different Energy Sources, 2007[a]

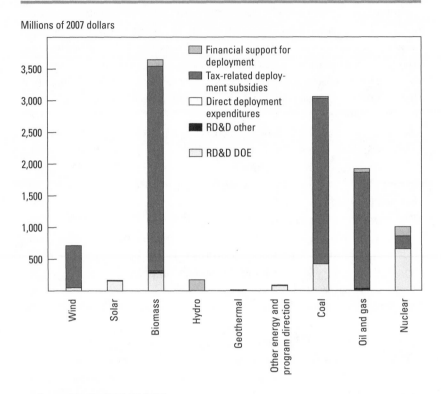

Millions of 2007 dollars

Source: Gallagher (2008); EIA (2008).
a. Other agencies are the U.S. Department of Agriculture, U.S. Geological Survey, and Defense Advanced Research Projects Agency. "Other energy" category includes other renewable energy programs (such as the landfill gas and international renewable energy programs) and renewable energy program direction costs.

—The fifth-largest expenditure ($0.7 billion) was the New Technology Credit (also known as production tax credit or PTC) for wind power. The wind PTC went into effect in 1994, and since its introduction overall wind capacity in the United States has grown from 2 GW to 17 GW in spite of the 1999, 2001, and 2003 PTC expirations, which led to three- to sevenfold drops in yearly capacity additions in the subsequent years.[69] The U.S. experience

69. American Wind Energy Association, "Wind Power Outlook 2007" (www.awea.org/pubs/documents/outlook_2007.pdf [January 2009]).

with wind PTC illustrates its effectiveness in promoting wind power deployment and the importance of predictable policies to encourage private sector investment.

One of the few reliable bodies of information combining public and private expenditures in RD&D, deployment subsidies, installed capacities, and changes in unit costs for a specific energy technology relates to PV systems in Japan.[70] In a seminal paper, Grübler and coworkers used the Japanese PV story to show that reductions in unit costs are a function of both RD&D and deployment and commercialization investments, that is, of suppliers and users of technology.[71] Hence a very important consideration for designing deployment policies is that the ETI system should be assessed as a whole.

Recommendations on Deployment Incentives

The U.S. government could make a difference by reducing significant barriers to the large-scale deployment of technologies that are relatively close to the market by

—providing investment certainty in the renewable energy arena by extending the eligibility of the PTC in five-year periods, instead of one- or two-year periods;

—enhancing the tax incentives for energy efficiency investments in the Energy Policy Act of 2005;

70. Watanabe, Wakabayashi, and Miyazawa (2000). The Japanese Ministry of Economy, Trade, and Industry (METI) has funded PV RD&D programs since 1974. A sharp increase in Japanese PV RD&D funding between 1979 and 1981, from $3.3 million to $56.7 million, stimulated an increase in private sector PV RD&D investment from approximately $6.5 million to $92.9 million within about a year (Watanabe, Wakabayashi, and Miyazawa 2000). METI's large increase in PV RD&D was accompanied by a PV cost buy-down program (a residential PV system dissemination subsidy), which was initiated in 1994 and administered through the New Energy Foundation. The size of the subsidy per system was gradually decreased—it was reduced by a factor of 7.5 between 1994 and 2001 and terminated in 2005—while the total yearly subsidy reached $200 million in 2001 (Jäger-Waldau 2003). The coordinated Japanese government action combining technology-push and market-pull incentives reduced PV costs from $164 per watt in 1974 to $6 per watt in 2006, and it turned Japan into a world leader in PV technology: in terms of existing PV capacity (over 1.7 GW in 2006), Japan is second only to Germany, and three out of the top five PV cell manufacturers by production capacity are Japanese. In January 2008, the Japanese prime minister announced a new goal of increasing PV capacity in Japan by a factor of 10 by 2020 and by a factor of 40 by 2030. This initiative requires ensuring that over 70 percent of newly built private buildings in Japan use solar energy. In addition, METI will start new subsidies for residential solar power generation in 2009. See Prime Minister of Japan and His Cabinet, "In Pursuit of 'Japan as a Low-Carbon Society.'"

71. Grübler, Nakićenović, and Victor (1999).

—directing greater resources to the commercialization of carbon capture and storage (CCS) and ensuring that CCS is included in any taxpayer-supported activity to develop coal-to-liquids technology;

—redirecting resources currently devoted to energy subsidies that are not effectively targeted (for example, VEETC and oil and gas tax credits) to programs that will promote the deployment of more promising options such as renewable power, cellulosic ethanol, clean high-quality diesel from organic waste, advanced nuclear waste disposal, more energy-efficient technologies in the industrial, transportation, and buildings sector, and fossil power with CCS technologies;

—providing targeted consumer and manufacturer incentives to encourage the domestic production, demonstration, and deployment of advanced automotive technologies; and

—addressing other barriers to the large-scale deployment of biofuels, including critical supporting infrastructures (gathering, distribution, and refueling systems) and compatible vehicle technologies.[72]

Standards-Based Policies

The United States has used two main types of standards to reduce energy consumption and promote the deployment of low-carbon technologies: energy efficiency (performance) standards and renewable portfolio standards (RPS).

Energy efficiency standards increase the average efficiency of equipment. Although it has been argued that it is not clear whether performance standards are able to stimulate improvements over the existing, most efficient technologies, or whether the same outcomes could have been reached at lower costs using market-based approaches such as cap-and-trade systems, performance standards can be very effective because they force people and firms to access information and replace older, less efficient technologies more rapidly than would otherwise be the case.[73] Corporate Average Fuel Economy (CAFE) standards have made, and will continue to make, the automobile fleet more fuel efficient.[74] The coordination of RD&D programs, a series of more stringent standards, and the Super Efficient Refrigerator Program have also

72. Most of these recommendations come from the 2007 National Commission on Energy Policy report (NCEP 2007).

73. Jaffe, Newell, and Stavins (2004).

74. The Energy Security and Independence Act of 2007 was the first federal statute to require an increase in fuel economy since 1975, calling for a CAFE standard of 35 miles per gallon by 2020—a 40 percent increase in efficiency.

Figure 5-7. State RPS Policies and Nonbinding Renewable Energy Goals[a]

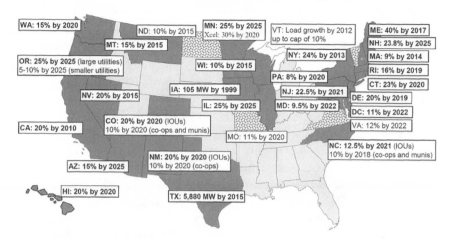

Source: Wiser and Barbose (2008).

a. IOUs, investor-owned utilities; Xcel, Xcel Energy, headquartered in Minneapolis, Minn.; munis, municipal electric companies.

resulted in dramatic increases in energy efficiency and cost-effectiveness in refrigerators (see figure 5-5). Fluorescent lighting systems and more efficient clothes washers are other examples of technologies that achieved an accelerated deployment through a combination of RD&D and energy efficiency standards.[75]

RPS policies establish numerical targets for renewable energy supply and impose them on retail electricity suppliers who can generate or purchase electricity from renewable sources. They have been increasingly used to promote renewable energy generation at the state level since the late 1990s, and as of December 2007, twenty-five states and Washington, D.C., had mandatory RPS policies (figure 5-7).[76] Regarding their impact on ETI, well-designed RPS policies encourage competition among renewable energy developers to meet targets in the least costly fashion, reduce market and cost barriers (in the sense that an RPS policy gives certainty about the existence and size of the renewable energy market), and might accelerate the large-scale deployment

75. David Hawkins, "Climate Change Technology and Policy," testimony before the Senate Committee on Commerce, Science, and Transportation, July 10, 2001 (www.nrdc.org/globalwarming/tdh0701.asp).

76. Once all mandatory RPS programs are implemented, they will apply to the entities supplying 46 percent of the retail electricity sales in the United States. See Wiser and Barbose (2008).

of technologies that are still not cost competitive. In 2007, 76 percent of all nonhydroelectric installed renewable energy capacity additions in the United States took place in states with RPS policies.

Recommendations on Standards-Based Policies

The standards-based recommendations listed below will complement the policy package to accelerate the deployment of more advanced energy technologies:

—Improve heavy-truck fuel economy standards and adopt efficiency standards for light-duty vehicle replacement tires.

—Ensure that DOE follows through on its commitment to issue new standards for more than thirty residential appliances and commercial equipment between 2006 and 2011.[77]

—Adopt a federal RPS to increase the share of renewable electricity to at least 15 percent by 2020.

—Expand the use of the Energy Star model for symbols and information campaigns to influence consumer behavior in the purchase of energy-efficient vehicles, building materials, and homes.[78]

Climate Policy: A Market-Based Program to Limit GHG Emissions

A well-designed, long-term, and predictable approach to penalizing emissions of heat-trapping gases—either a cap-and-trade system or a tax—would minimize the marketwide cost of achieving the needed reductions in these emissions. Because firms would have flexibility regarding whether to pay for emissions or reduce them, the market would tend to find the reductions that were cheaper than the permit price or tax. In addition, either the cap-and-trade or the tax approach would send price signals that would incentivize firms to invest in the development and deployment of less expensive low- and no-carbon technologies.[79]

By itself, however, even a well-designed cap-and-trade or tax approach will not be able stimulate all the ETI that is needed because of the market failures associated with the R&D enterprise (for example, knowledge spillovers).[80]

---

77. As of November 2007, it had adopted new standards for two appliances. See DOE, "DOE Increases Energy Efficiency Standards for Residential Furnaces and Boilers," press release, November 19, 2007 (www.energy.gov/news/5743.htm).

78. This action should be complemented by encouraging the publicity efforts of active local program sponsors, as it has been shown that such efforts increase the recognition, understanding, and influence of the label (U.S. Environmental Protection Agency 2008).

79. Stavins (2008).

80. Equally, ERD&D alone will not result in the needed reduction in GHG emissions without climate change policies limiting emissions. See Edmonds and others (2004).

Furthermore, any near-term market-based climate regulation is unlikely to be sufficient (on its own) to overcome the barriers to the deployment of some of the most needed GHG-mitigation technologies, such as CCS, because the political process required to reach agreement on climate legislation is likely to result in too low an allowance price or tax . Supplementary measures to close this gap (such as allocation of extra emissions allowances to companies that deploy CCS) will be required, along with the other kinds of ETI policies that have been discussed here.[81]

Recommendations for Climate Policy

Congress should pass promptly a mandatory, economy-wide approach to penalizing GHG emissions, based on either emission taxes or emissions limits implemented through tradable permits, in either case becoming more stringent with time.[82] The approach should be designed to achieve at least a 15 percent reduction in U.S. GHG emissions below 2006 levels by 2030, consistent with the much deeper reduction that will be needed by 2040, 2050, and beyond if the world is to have a reasonable chance of avoiding unmanageable changes in global climate.[83] The country and the world cannot afford to wait for a "perfect" approach to GHG emissions reductions, that is, one that satisfies all stakeholders. Instead, the GHG-reduction policy should have built-in flexibilities to adapt to future economic, environmental, and technological information.

Conclusion

A substantial acceleration of the pace of energy technology innovation in the United States and around the world is an absolutely necessary—even if not sufficient—condition for meeting in a timely way the massive challenges associated with the increasing demand for reliable and affordable energy, the need to reduce the economic and security liabilities of overdependence on oil, and the danger of an unmanageable degree of global climate disruption driven by fossil fuel–derived greenhouse gas emissions. The United States, as the world's largest consumer of oil and, with China, one of the two largest

81. NCEP (2007).

82. State governments have also moved ahead of the federal government in this area. The Regional Greenhouse Gas Initiative—a cap-and-trade program that includes ten northeastern states—and the California Greenhouse Gas Solutions Act (AB32) are scheduled to begin cutting emissions in 2015 and 2012, respectively.

83. UN Scientific Expert Group on Climate Change and Sustainable Development (2007).

emitters of carbon dioxide from fossil fuels, has an obligation to play a leadership role in bringing about the needed acceleration of ETI, not only in this country but around the world. And it has the capacity to do so. But it has been falling far short, in this domain, of the effort that is needed.

It is not difficult, moreover, to identify the principal elements of a set of policies that are capable of at least beginning to harness the collective capacities of the public and private sectors in this country to perform this immensely important task. We have tried to lay out here much of what would be required. It is a large agenda, but with leadership from the White House and Congress it could all be done. It must be hoped that the new administration and the new Congress will finally provide that leadership because there is no more time left to waste.

References

Anderman, Menahem. 2007. *The 2007/8 Advanced Automotive Battery and Ultracapacitor Industry Report.* Oregon House, Calif.: Advanced Automotive Batteries.

Audretsch, David B., ed. 2006. *Entrepreneurship, Innovation and Economic Growth.* Northampton, Mass.: Edward Elgar.

Baldi, Stéphane, and others. 2007. *Highlights from PISA 2006: Performance of U.S. 15-Year-Old Students in Science and Mathematics Literacy and in an International Context.* NCES 2008–016. U.S. Department of Education, National Center for Education Statistics.

Boyle, Rohan, and others. 2008. *Global Trends in Sustainable Energy Investment 2008: Analysis of Trends and Issues in the Financing of Renewable Energy and Energy Efficiency.* Paris: United Nations Environment Program, Division of Technology, Industry and Economics.

Brooks, Harvey. 1967. "Applied Science and Technological Progress." *Science* 156, no. 3783: 1706–12.

Brown, Marilyn A., Frank Southworth, and Therese K. Stovall. 2005. *Towards a Climate-Friendly Built Environment.* Arlington, Va.: Pew Center on Global Climate Change.

Brown, Marilyn A., and others. 2007. "Carbon Lock-In: Barriers to Deploying Climate Change Mitigation Technologies." Oak Ridge, Tenn.: Oak Ridge National Laboratory.

Bunn, Matthew. 2008. "Safety, Security, Safeguards: Enabling Nuclear Energy Growth." Presentation at the Global Nuclear Future Workshop, American Academy of Arts and Sciences, Cambridge, Mass., May 5–6.

Commission of the European Communities. 2007. "A European Strategic Energy Technology Plan (SET-Plan): 'Towards a Low Carbon Future.'" SEC(2007) 1508–11. Brussels: European Union.

David, Paul A., Bronwyn H. Hall, and Andrew A. Toole. 1999. "Is Public R&D a Complement or Substitute for Private R&D? A Review of the Econometric Evidence." *Research Policy* 29, no. 4-5: 497–529.

Deutch, John, and Ernest Moniz. 2003. *The Future of Nuclear Power: An Interdisciplinary MIT Study.* Massachusetts Institute of Technology.

Dooley, James J. 1999. "Energy R&D in the United States." PNNL-12188. Richland, Wash.: Pacific Northwest National Laboratory.

Edmonds, Jae, and others. 2004. "Stabilization of CO_2 in a B2 World: Insights on the Roles of Carbon Capture and Disposal, Hydrogen, and Transportation Technologies." *Energy Economics* 26, no. 4: 517–37.

Gallagher, Kelly Sims. 2008. "DOE Budget Authority for Energy Research, Development, and Demonstration Database." Harvard, Kennedy School of Government, Energy Technology Innovation Policy Group.

Gallagher, Kelly Sims, John P. Holdren, and Ambuj D. Sagar. 2006. "Energy-Technology Innovation." *Annual Review of Environmental Resources* 31 (November): 193–237.

Gallagher, Kelly Sims, Robert Frosch, and John P. Holdren. 2004. "Management of Energy-Technology Innovation Activities at the Department of Energy." Technical Appendix. Washington: National Commission on Energy Policy.

Geller, Howard S., and David B. Goldstein. 1999. "Equipment Efficiency Standards: Mitigating Global Climate Change at a Profit." *Physics and Society* 28, no. 2: 4.

Grübler, Arnulf, Nebojša Nakićenović, and David G. Victor. 1999. "Dynamics of Energy Technologies and Global Change." *Energy Policy* 27, no. 5: 247–80.

Hightower, Mike, and Suzanne A. Pierce. 2008. "The Energy Challenge." *Nature* 452, no. 20: 285–86.

Holdren, John P. 2006. "The Energy Innovation Imperative: Addressing Oil Dependence, Climate Change, and Other 21st Century Energy Challenges." *Innovations: Technology, Governance, Globalization* 1 (Spring): 3–23.

Intergovernmental Panel on Climate Change (IPCC). 2007a. *Climate Change 2007: Synthesis Report.*

———. Working Group II. 2007b. "Summary for Policymakers." In *Climate Change 2007: Impacts, Adaptation and Vulnerability Contribution of Working Group II to the Fourth Assessment Report of the Intergovernmental Panel on Climate Change.* Cambridge University Press.

International Energy Agency 2007. *World Energy Outlook 2007: China and India Insights.*

Jaffe, Adam B., Richard G. Newell, and Robert N. Stavins. 2004. "Technology Policy for Energy and the Environment." In *Innovation Policy and the Economy*, vol. 4, edited by Adam B. Jaffe, Josh Lerner, and Scott Stern, pp. 35–68. MIT Press.

Jäger-Waldau, Arnulf. 2003. *PV Status Report 2003. Research, Solar Cell Production, and Market Implementation in Japan, USA, and the European Union.* EUR 20850EN. Ispra, Italy: European Commission, Directorate-General Joint Research Center.

Kammen, Daniel M., Kamal Kappadia, and Matthias Fripp. 2004. *Putting Renewables to Work: How Many Jobs Can the Clean Energy Industry Create?* Report of the Renewable and Appropriate Energy Laboratory, University of California–Berkeley.

Kuenzi, Jeffrey K. 2008. "Science, Technology, Engineering, and Mathematics (STEM) Education: Background, Federal Policy, and Legislative Action." Report RL33434. Congressional Research Service, Library of Congress.

Makower, Joel, Ron Pernick, and Clint Wilder. 2008. "Clean Energy Trends 2008." San Francisco, Calif.: Clean Edge (March).

Metcalf, Gilbert E. 2008. "Using Tax Expenditures to Achieve Energy Policy Goals." Working Paper 13753. Cambridge, Mass.: National Bureau of Economic Research.

Moran, Robert. 2007. "Photovoltaics: Global Markets and Technologies." Report EGY014F. Wellesley, Mass.: BCC Research.

National Academy of Sciences. Committee on Prospering in the Global Economy of the 21st Century. 2005. *Rising Above the Gathering Storm: Energizing and Employing America for a Brighter Economic Future.* Washington: National Academy Press.

National Commission on Energy Policy (NCEP). 2004. *Ending the Energy Stalemate. A Bipartisan Strategy to Meet America's Energy Challenges.* Washington.

———. 2007. "Energy Policy Recommendations to the President and the 110th Congress."

National Research Council (NRC). Board on Energy and Environmental Systems. 2001. *Energy Research at DOE: Was It Worth It? Energy Efficiency and Fossil Energy Research 1978 to 2000.* Washington: National Academy Press.

National Science Foundation. 2008. *Science and Engineering Indicators 2008.*

Nemet, Gregory F., and Daniel M. Kammen. 2007. "U.S. Energy Research and Development: Declining Investment, Increasing Need, and the Feasibility of Expansion." *Energy Policy* 35, no.1: 746–55.

Ogden, Peter, John Podesta, and John Deutch. 2008. "A New Strategy to Spur Energy Innovation." *Issues in Science and Technology* 24 (Winter): 35–44.

Paltsev, Sergey, and others. 2007. "Assessment of U.S. Cap-and-Trade Proposals." Report 146. MIT, Joint Program on the Science and Policy of Global Change.

Parmesan, Camille, and Gary Yohe. 2003. "A Globally Coherent Fingerprint of Climate Change Impacts across Natural Systems." *Nature* 421(January 2): 37–42.

President's Committee of Advisors on Science and Technology (PCAST). 1997. *Report to the President on Federal Energy Research and Development for the Challenges of the Twenty-First Century.* Washington.

———. 1999. *Powerful Partnerships: The Federal Role in International Cooperation on Energy Innovation.* Executive Office of the President, Office of Science and Technology Policy.

Renewable Energy Policy Network for the Twenty-First Century (REN21). 2008. *Renewables 2007: Global Status Report.* Paris: REN21 Secretariat, and Washington: Worldwatch Institute.

Runci, Paul. 2005. "Energy R&D Investment Patterns in IEA Countries: An Update." Paper PNWD-3581. University of Maryland, Joint Global Change Research Institute.

Schock, Robert, and others. 1999. "How Much Is Energy Research and Development Worth as Insurance?" *Annual Review of Energy and Environment* 24: 487–512.

Stavins, Robert N. 2008. "Addressing Climate Change with a Comprehensive U.S. Cap-and-Trade System." *Oxford Review of Economic Policy* 24, no. 2: 298–321.

UN Development Program. 2000. *World Energy Assessment: Energy and the Challenge of Sustainability.* New York.

UN Scientific Expert Group on Climate Change and Sustainable Development. 2007. *Confronting Climate Change: Avoiding the Unmanageable and Managing the Unavoidable.* Washington: United Nations Foundation and Sigma Xi.

U.S. Department of Energy (DOE). 2000. "A Decade of Success, Office of Energy Efficiency and Renewable Energy." DOE/EE-0213.

U.S. Energy Information Administration (EIA). Office of Coal, Nuclear, Electric, and Alternate Fuels. 2008. *Federal Financial Interventions and Subsidies in Energy Markets 2007.*

U.S. Environmental Protection Agency. Office of Air and Radiation. 2008. *National Awareness of ENERGY STAR® for 2007: Analysis of 2007 CEE Household Survey.*

U.S. General Accounting Office. 1996. "Department of Energy: Opportunity to Improve Management of Major System Acquisitions."

Watanabe, Chihiro, Kouji Wakabayashi, and Toshinori Miyazawa. 2000. "Industrial Dynamism and the Creation of a 'Virtuous Cycle' between R&D, Market Growth and Price Reduction. The Case of Photovoltaic Power Generation (PV) Development in Japan." *Technovation* 20, no. 6: 299–312.

Wiser, Ryan, and Galen Barbose. 2008. "Renewables Portfolio Standards in the United States. A Status Report with Data through 2007." LBNL-154E. Berkeley, Calif.: Lawrence Berkeley National Laboratory.

six
Electricity Market Structure and Infrastructure

William W. Hogan

Infrastructure investment is a common focus of energy policies proposed for the United States. Initiatives to improve energy security, meet growing demand, or address climate change and transform the structure of energy systems all anticipate major infrastructure investment. Long lead times and critical mass requirements for these investments present chicken-and-egg dilemmas. Without the necessary infrastructure investment, energy policy cannot take effect. And without sound policy, the right infrastructure will not appear. Acting in time to provide workable policies for infrastructure investment requires a framework for decisionmaking that identifies who decides and how choices should be made.

Infrastructure investment for the electricity sector in generating plants, transmission lines, distribution wires, control systems, metering and end-use devices is an important part of the larger picture. The electricity sector is capital intensive and under any reasonable forecast requires substantial investments. Choosing which investment to make, and who should pay, can be controversial. The controversies are compounded by the continuing transition in the electricity sector to restructure the balance between markets and regulation, between incentives and central planning, and between risk and reward.

The purpose here is to outline the major investment policy challenges for the electricity sector in light of this restructuring transition. Notably, there is an important role for government, through regulation and centralized coordination, and for markets, through incentives and innovation. Implementing a workable balance between the role of government and the contribution of markets requires looking ahead to act now to reinforce rather than undermine long-term goals. Special features of the electricity system interact with more general energy problems to make this balancing act a formidable challenge. There are several strands that have to work together, and actions now must be forward-looking to avoid unintended consequences that could unravel the larger fabric.

Investment and Uncertainty

The North American Electric Reliability Corporation (NERC) provides annual assessments of the outlook for reliability of the electricity system. The assessments provide a broad overview of the electricity infrastructure and expected investment. According to NERC's most recent assessment, the United States will add 123 gigawatts (GW) of new capacity between 2007 and 2016.[1] This compares, for example, with the 100 GW capacity of the existing nuclear fleet. However, the expectation is that most of this new capacity would come from coal and natural gas with a small contribution from other technologies. In addition, NERC anticipates over 14,000 circuit miles of additions to the high-voltage power grid.[2] These would be substantial investments, but even with this outlook, NERC reports that "industry professionals ranked aging infrastructure and limited new construction as the number one challenge to reliability—both in likelihood of occurrence and potential severity."[3]

From one perspective, this pace of investment looks quite manageable. Total installed capacity of generating plants is now more than 1,000 GW, so the expansion of 123 GW over a decade is only about 12 percent. This is less than the 281 GW of new capacity investment, primarily in natural gas plants, that entered the system over the seven years from 1998 to 2005.[4] Similarly, the expanded transmission investment is less than 10 percent of the existing 163,000 circuit miles. In the aggregate, therefore, the investment needed to meet the expected growth in demand seems well within our capabilities.

However, the aggregate figures hide regional differences that give rise to a need for further investment. "Areas of the most concern include WECC-Canada, California, Rocky Mountain States, New England, Texas, Southwest

1. NERC (2007, p. 10).
2. Ibid., p. 18.
3. Ibid., p. 19.
4. NERC (2006, p. 31).

and the Midwest."[5] Furthermore, the choice of fuel and technology presents significant difficulties in deciding on the portfolio of investments. "The unique characteristics and attributes of renewables require special considerations for planning. For example, they are often remotely located, requiring significant transmission links often over challenging terrain."[6] A national policy to transform the electricity sector to address the problems of climate change would compound these difficulties.

In addition to the regional variation, there are the substantial uncertainties that cloud investment decisions. Electricity-generating plants differ in important ways that complicate the selection of technology. For instance, in its development of the Annual Energy Outlook (AEO), the U.S. Energy Information Administration (EIA) considers generation technologies that range in initial capital cost from a low of $450 per kilowatt (kW) for advanced combustion turbines, to $1,434 per kW for scrubbed coal, and $2,143 per kW for advanced nuclear power generators. Similarly, the expected (and optimistic) construction times are two, four, and six years, respectively.[7] The capital cost numbers illustrate the range of choices but substantially understate current estimates of the level of such costs at more than double the EIA planning assumptions, reinforcing the point of considerable uncertainty.[8] When construction time is added to planning and permitting, the lead times could extend out for a decade. On the demand side, the array of possible technologies and usage patterns is highly diverse, and technology is changing rapidly. This creates many other opportunities to exploit, with a great variety of costs and impacts.

If these figures were known with confidence and the future were easily predicted, this great variety in costs and planning horizons would not present much of a problem. Small errors would be temporary and overtaken by growth in demand. However, the reality is that there is great uncertainty that confounds the forecasts and infrastructure investment planning.

The connection between investment and uncertainty arises because of the long lead times and long lives of infrastructure investments. Were the uncertainty simply random variation, the cost would be clear, and the prescriptions would call for diversification, hedging, and more but smaller investments. However, the uncertainty that affects the energy system is not just

5. NERC (2007, p. 10).
6. Ibid., p. 13.
7. EIA (2008, p. 79).
8. Federal Energy Regulatory Commission (FERC), "Increasing Costs in Electric Markets," staff report, June 19, 2008 (www.ferc.gov/legal/staff-reports/06-19-08-cost-electric.pdf).

random variation. The rapidly changing policy and economic system, with the constant flux of technology innovation, produces information about opportunities and risks that call out for a different way of making investment decisions. The balance shifts away from an emphasis on central planning and the judgments of the few to the more dispersed wisdom of highly motivated crowds that are bearing the risks and reaping the rewards. A challenge in policy for the electricity system is to craft and sustain this change in balance given the large uncertainties ahead.

One way to develop an appreciation for the degree of uncertainty is to look at the record from the past. The EIA has been publishing its annual outlooks for many years. The methodology and knowledge applied have changed with accumulating experience, but there is enough stability in the people and processes to give some reason to compare the results of the forecasts and the actual events. The EIA publishes regular retrospectives, and the forecasting horizon for which we have the most data is a seven-year projection. Seven years is not a long time in the electricity infrastructure investment cycle, with many decisions that depend on outcomes over much longer horizons.

Figures 6-1 and 6-2 summarize the EIA record for the seven-year forecast error as a percent for key prices and quantities over almost two decades of publication experience. For those familiar with these matters, the forecast errors are not surprising, and perhaps less than expected. The EIA is highly professional and serious, but these errors over a relatively short horizon for prices are considerable and reflect the difficulty of forecasting, as made evident by the unanticipated rise in prices in 2007–08 when oil went to $140 per barrel. In the quantity forecasts, there is more stability, but the range of errors is still large and of the same order of magnitude as the forecast infrastructure investment described in the NERC assessments.

A high degree of uncertainty arises in part because of the uncertainty in energy policy going forward, over which the government has some control, and in part because of the inherent unpredictability of economic events.

Concerns with policy uncertainty are easy to find. For example, the pressure to address the challenge of climate change and carbon emissions is building, with the expectation that there will be controls and a price on carbon. Since the whole purpose of pricing carbon and controlling emissions is to dramatically change the portfolio of investments, it should not be surprising that the policy uncertainty complicates investment infrastructure planning.

A broad analysis of the impacts of policy uncertainty can be found in a recent study conducted by the International Energy Agency (IEA) reviewing

Figure 6-1. Seven-Year EIA Price Forecast Errors, 1982–2000[a]

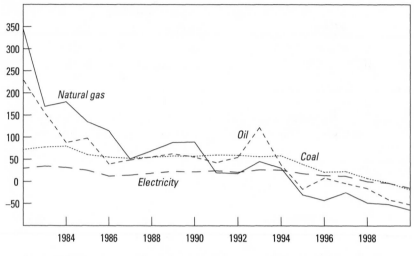

Percent predicted less than actual

Source: EIA (2007).

the experience in developing electricity markets across IEA member countries.[9] The comparative experience shows ample evidence that with the right conditions, the necessary investments will be made. "Experience to date shows that, with the right incentives and with a stable investment climate, investors are responsive to the needs for new generation capacity. When signals are undistorted in effectively liberalised markets and companies have incentives to compete, investors respond to market signals and have so far added new capacity on time."[10] However, the same analysis emphasizes the importance of dealing with the high level of uncertainty confounding investment decisions.

The IEA's extensive comparison across countries and its advice to governments are relevant to the U.S. government and, especially, the Federal Energy Regulatory Commission (FERC), with its responsibility to oversee wholesale

9. The IEA member countries are Australia, Austria, Belgium, Canada, the Czech Republic, Denmark, Finland, France, Germany, Greece, Hungary, Ireland, Italy, Japan, the Republic of Korea, Luxembourg, the Netherlands, New Zealand, Norway, Portugal, Spain, Sweden, Switzerland, Turkey, the United Kingdom, and the United States.

10. IEA (2007, p. 13).

Figure 6-2. Seven-Year EIA Quantity Forecast Errors, 1982–2000[a]

Percent predicted less than actual

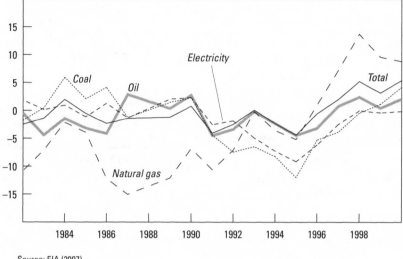

Source: EIA (2007).

electricity markets. The IEA report provides a lengthy analysis to support "key messages" that summarize the conclusions, as shown in box 6-1.

Government can do something to reduce, if not eliminate, policy uncertainty. There is less that government can do to address the inherent uncertainty in economic events. Here, the issues center on who should decide and how decisions should be made—through planning or through markets—and how risks should be allocated—to captive customers or to market investors—to best address the many choices and the complications induced by the uncertainties.

Except for the climate change policy decision, under current U.S. laws and regulations, FERC has primary responsibility for setting and regulating the policies and practices needed to follow through on the IEA recommendations. Translating the general principles into practical rules and policies presents a major agenda for FERC.

A central problem for infrastructure investment is dealing with uncertainty and allocating the associated risks. The uncertainty is so large that it may be the most important variable, more than the forecast central values. In addressing this decision and regulatory framework, a review of the electricity

Box 6-1. IEA Key Messages for Government Role in Energy Markets

Governments must ensure a stable and competitive investment framework that sufficiently rewards adequate investments in a timely manner. Considerable investment in new power generation will be required over the next decade to meet increasing demand and replace ageing generation units. Current trends suggest a significant risk of underinvestment. Long project lead times and high investment costs, particularly for large base-load units, create a need for government action to reduce uncertainty in the very near term. Efficient use of existing resources is particularly important at this stage, as it allows for lower margins and buys time to meet investment requirements.

. . .

Governments urgently need to reduce investment risks by giving firmer and more long-term direction on climate change abatement policies. Putting a price on greenhouse gas emissions is an effective way to internalize the costs of climate change. Direct financial support for specific technologies, such as renewables and nuclear, should be done at the lowest cost and with market-compatible instruments. Market-based instruments, such as tradeable obligations systems, have many advantages; direct subsidies, such as tax credits, can also be implemented in ways that are compatible with competitive markets. Nuclear power will only play a more important role in climate change abatement if governments in countries where nuclear power is accepted play a stronger role in facilitating private investment.

. . .

Governments should pursue the benefits of competitive markets to allow for more efficient and more transparent management of investment risks. Competition in well-designed and effectively liberalised markets creates incentives for efficient use of resources and investments in power generation. However, in order to deliver its anticipated benefits, liberalisation requires whole-hearted implementation and long-term commitment by governments. Competition cannot always stand alone. When necessary, governments should pursue intervention in ways that complement the market and facilitate its functioning.

. . .

Governments need to ensure that independent regulators and system operators establish transparent market rules that are clear, coherent and fair. Transmission system operators hold the key to competitive electricity markets and must be effectively separated from generation and retail supply. Unified regulation and unified system operation should be pursued as tools to facilitate dynamic trade across borders and efficient sharing of reserves.

. . .

Governments must refrain from price caps and other distorting market interventions. Wholesale electricity prices are inherently volatile and price spikes are an integral part of a competitive market. Price caps, regulated tariffs that undercut market prices, and direct market intervention seriously undermine market confidence, jeopardizing efficiency and reliability. Governments can best address systemic market power abuse by: improving market design; strengthening competition law and competition regulators; and diluting the dominance of large players. Demand response constitutes an essential but still poorly exploited resource, and must receive specific attention in the development of market design and regulation. Increased installation of better metering and control equipment could considerably strengthen potential demand response resources. Capacity measures may be necessary if price caps are imposed. However, they are not a preferred solution to address market power and can easily become a barrier for the development of a robust market.

. . .

Governments must implement clearer and more efficient procedures for approval of new electricity infrastructure. Delays caused by slow licensing and inefficient approval procedures frustrate markets, and are serious barriers to timely investment. Governments must rebalance competing interests in favor of new electricity system infrastructure and offer clearer and more efficient approval procedures, preferably centered on one approval body. Timelines for approval processes must be clear and established in advance. Fast and efficient licensing is particularly important for new nuclear power plants which face very high risks as well as long planning and approval process. Early public debate is essential for the acceptance of necessary new infrastructure.

Source: Excerpted from International Energy Agency (2007, pp. 15–25).

restructuring transition provides the setting for translating the key messages from the IEA that apply to the actions of government in the United States to advance the framework for infrastructure investment.

Electricity in Transition

The electricity sector has been experiencing a major restructuring of its infra- structure, institutions, and choices. The old model was built around the prin- ciple of monopoly franchise with a regulatory structure that assumed a high degree of vertical integration and a reasonable degree of certainty. For decades following the consolidations associated with Samuel Insull in the early twen- tieth century, electricity companies built and operated their own infrastruc- ture in the form of generating plants for producing power, high-voltage trans- mission lines for moving power, and distribution wires for connecting with customer loads. Customer service was for full requirements, meaning that all electricity power and associated services were provided in a bundle by the elec- tricity company. Customers had little or no choice as to who supplied their power. In exchange for this monopoly franchise, under the Insull framework, electricity companies accepted extensive price and service regulation that con- trolled the rates charged to customers to reflect regulators' judgments about reasonable cost recovery and appropriate returns on capital invested.

The electricity system was never quite as simple as this stylized outline. There were always variations in the form of municipal and public power companies that differed from the investor-owned utilities. But in its essence, the combination of natural monopoly, vertical integration, full requirements service, and rate regulation provided the template for industry organization. This framework had important implications for infrastructure investment. Primary responsibility for deciding on the pace and scope of infrastructure investment rested with the vertically integrated utility, which had an obliga- tion to serve the full requirements of its customers. The decisions made by management built on a central planning process that attempted to look long term and invest ahead of growing demand for electricity. Regulators would review investments after the fact to judge the reasonableness of the expendi- tures and allow the costs of the new infrastructure to be included in the "rate base" that determined the prices charged to the final customers.

This system worked extremely well for many decades, with rapidly grow- ing use of electricity coupled with declining average costs and rates. For the four decades before 1970, nominal electricity prices were declining or roughly level, and real prices were falling. Demand growth was large and predictable, averaging over 8 percent a year.[11] In retrospect, this happy outcome can be attributed to the combined effects of technology improvements, exploitation of economies of scale and scope, and relatively predictable costs and technology. Regulators were in the position of approving expanded investments that would lower average costs and produce stable nominal and generally declining real prices for electricity.

As summarized in figure 6-3, the story changed in the 1970s. There appeared to be an exhaustion of the economies of scale that made new investment cheaper. Fuel prices rose in response to the oil price shocks and related energy crises that began with the first Arab oil embargo in 1973. Nuclear power plant costs started to rise, and there were growing delays in plant completion, adding even greater costs and putting more pressure on the electricity industry. The results produced a reversal in the long-term trend in electricity prices. Both nominal and real electricity prices started to climb, with real electricity prices up more than 50 percent by the early 1980s. The nuclear accident at Three Mile Island in 1979 and the 1983 default on the bonds of the Washington Public Power Supply System foreshadowed even further technical and financial problems that would unsettle the previously stable electric power system.[12] Thus began a period of turmoil and extended transition for customers, electricity companies, and their regulators.

These events had two important implications for the framework for infrastructure investment. First, the easy days of adding new investments to the rate base that lowered average costs were replaced by the need for regulators to approve increases in electricity rates, and this produced a new era of challenges and cost disallowances. No longer could the utility company be sure that the costs of major investments would be fully recognized and recovered. The resulting financial distress for many companies fundamentally changed the incentives and the willingness to invest in new infrastructure.

Second, the locus of decisions began to change. Before, planners at the electric utility companies made the choices for the generation, transmission, and distribution portfolio of investments. The role of government centered on regulation of the resulting prices charged to the franchise customers. But legislation such as the Public Utility Regulatory Policies Act of 1978 (PURPA)

---

11. Rustebakke (1983, p. 4).

12. On March 28, 1979, the Three Mile Island Unit 2 nuclear power plant near Middletown, Pennsylvania, suffered a partial core melt (U.S. Nuclear Regulatory Commission 1979). In 1983 Washington Public Power Supply System defaulted on $2.25 billion of bonds due to inability to complete five nuclear reactors. "It was the largest municipal bond default in U.S. history" (Myhra 1984, pp. 1–2).

Figure 6-3. U.S. Average Electricity Prices, 1960–2007

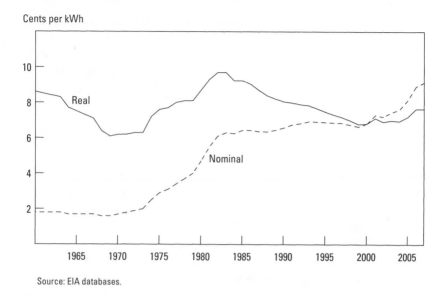

Source: EIA databases.

began a gradual change in this division of responsibilities.[13] Under section 210 of PURPA, electric utilities were required to purchase power from "qualifying facilities" (QFs) at a price based on an administrative estimate of the utility's avoided costs. Although seen initially as a modest intervention to give a boost to the small power plants and cogenerators that made up the QFs, the PURPA provisions and related initiatives in New York and California had a profound effect. As intended, the response from the competitive new entrants in small power and cogeneration demonstrated that they were more than capable of building and operating power plants. As not intended, the experience demonstrated that government planners were fully capable of selecting high-cost new investment that compounded the problems of the industry.[14] The planners' avoided cost estimates were so high that they produced both

13. 16 U.S.C. 2601 et seq.

14. In 1981 New York law required payment of six cents per kilowatt-hour ($60 per MWh) for QF power. See N.Y. Public Service Law sec. 66-c.1. In California the QF "standard offer" solicitations at avoided costs were so oversubscribed that the California regulators sought coordinated procurement through the "Biennial Resource Plan Update" (BRPU), which required utilities to put their planned new generation out to bid. In the end, the regulators never approved new plant construction in the BRPU proceeding. See Southern California Edison Company and others (1995), 70 FERC para. 61, 215, at p. 61,677. The collapse of the BRPU process played a prominent role in the move to reform electricity regulation in California.

too much new supply and too great an increase in total costs. By the early 1990s, the first phase of the electricity transition had solidified an interest in finding some other way to organize the industry, drive decisions, assign risks, and govern expansion of the electricity infrastructure.

A new approach came on the scene following from the experience in other industries and other countries. The natural monopoly argument, at least for electricity generation, was overtaken by events. With no new economies of scale, and with the growth of the QFs, it became plausible to argue that electricity generation need not be treated as a natural monopoly. Opening the electricity market to competition seemed possible and would provide an alternative means to monopoly and regulation to enforce efficient investment and operating decisions. Although this argument applied directly to wholesale competition and the production of bulk power supplies, similar arguments were applied to retail competition and creation of customer choice with unbundled energy services.[15]

The new idea was that like airlines, trucks, and natural gas, complete regulation of vertically integrated monopoly franchise electricity operations was not necessary, and a better way would be to shift more toward relying on competition in electricity markets. The next phase in the electricity transition began at the federal level with the Energy Policy Act of 1992 (EPAct92) and the introduction of wholesale competition.[16] Although EPAct92 explicitly excluded federal preemption to extend competitive markets to retail customers, states from California to Massachusetts undertook various initiatives to unbundle generation, transmission, and retail supply in parallel with federal actions to expand competition in the wholesale electricity market. The provisions of EPAct92 had profound effects. The law expanded the scope of QFs by creating a new class of exempt wholesale generators— essentially power producers that could be either independent or affiliated with traditional utilities but would be spared the usual restrictions under the regulations for utility holding companies.[17]

Most important, acting under EPAct92 authority, FERC required existing electric utilities to eliminate undue discrimination and give third parties access to their transmission systems in order to facilitate wholesale trading and com-

---

15. EIA (2000).

16. P.L. 102-486.

17. Public Utility Holding Company Act of 1935, P.L. 74-333 (PUHCA). The law provided for regulation under the Securities and Exchange Commission and was an earlier reform designed to restrict the activities of utility holding companies. PUHCA was repealed with the Energy Policy Act of 2005.

petition. Since in each region there is only one interconnected grid, it was clear that transmission services retained the features of a natural monopoly. Competition in generation and supply could not work without access to the grid and the ability to arrange transactions between producers and customers. Support of wholesale electricity market competition and transmission open access became national policy. However, implementing the abstract notions involved more than was anticipated.

Coordination for Competition

Passage of EPAct92 launched an extensive process of analysis and new regulatory initiatives to translate the abstract notions of open access and support of competitive markets into a workable set of new arrangements that would govern investment and operations. The largely unexpected result was that provisions requiring open access to the transmission grid to support wholesale market competition would fundamentally transform the regulatory and institutional structure of the electricity system.

Unbundling with eventual sale or separation of generating plants owned by existing electric utilities was simple in concept but difficult in practice. After the reversal of the increase in other fuel prices in the 1980s and the widespread excess capacity, in part because of the entry of QFs and reduced growth in electricity demand, the perception in many parts of the country was that the existing fleet of generating plants was uneconomic. A simple jump to open competition would drive down wholesale prices and leave the existing assets economically "stranded." This is a long story. During the days after passage of EPAct 92, most policy discussion was about who would bear the costs of the stranded assets. In effect, although with different rules applied in nearly every state that undertook a major restructuring, the result was that the costs were shared, with the greater portion born by customers. Whatever the merits of the sharing of stranded costs, the lesson for today is less about the allocations to different groups and more about the form of that allocation. In important cases, from California to Maryland, the structure of the rate deal was a long-term fixed price arrangement that was disconnected from the evolving market. Later, when prices in the market rose, typified by the electricity crisis in California in 2000–01, the structure of the rate deal was unsustainable. The political problems that followed illustrated the dangers of constructing an imbalance between markets and regulation. The unintended consequences were severe and almost fatal for the restructuring policy.

The ubiquitous and poorly understood role of transmission system operations presented an unusual and unexpected challenge. Unbundling of generation seemed conceptually simple and straightforward. Less obvious was how to provide open access to the transmission grid. The states were responsible for the most important initiatives restructuring the ownership and reallocating the risks for generation infrastructure. However, the mandates of EPAct92 gave the federal government responsibility to define the rules of transmission open access, with the bulk of the activity at FERC. There began an odyssey that was still incomplete more than a decade later. Through an extensive round of hearings and rulemaking processes, FERC examined most of the major issues that would determine the rules of open access. The core problem was deeply rooted in the nature of the transmission grid. As is often the case in policy development, a commonly held belief turned out to be seriously untrue. In this case, the false assumption was that there was a simple way to define a path in the transmission network along which power could be delivered from customer to generator, the so-called contract path. If this were true, setting up the rules for open access wholesale market competition would have been relatively simple, possibly emulating the rules that FERC had previously developed in the natural gas market. But if it were not true, FERC would need something else that might require a more radical change in institutions.

In its landmark Orders 888 and 889, promulgated in 1996, FERC produced a careful documentation and analysis of the dilemma.[18] As explained in Order 888, the "contract path" was a fiction at odds with actual operation of the transmission grid, where power actually flowed on every parallel path and every individual transaction could have a material affect on other transactions. However, FERC could see no available alternative and proceeded to adopt open access rules based on the fictional foundation of the contract path.[19]

When there is a vertically integrated monopoly, it matters only a little when the explicit rules for one part of the vertical chain do not conform to reality because the monopoly has little incentive to act in ways that conform to the rules but are counterproductive in the whole. However, in the unbundled system without full vertical integration, the whole point of the system is for many different decisionmakers to act in response to the incentives in the sector, advancing the performance of the whole while seeking their own balance of profit and risk. It was clear that Order 888 and the contract path

18. FERC (1996a, 1996b).
19. For further discussion, see Hogan (2002).

model would create chaos in a true open access market. In the event, NERC immediately implemented ad hoc transmission loading relief measures to undo the Order 888 schedules, but the NERC rules were widely recognized as inefficient.[20]

The response across the country followed two basic approaches and divided the country into two types of electricity systems. For most of the country, there was a major effort to organize markets in a completely different way. The basic idea was that a large region would recognize the strong interactions among the participants and create an independent system operator (ISO) or a regional transmission organization (RTO) to handle all of the real-time and some of the longer-term activities in providing transmission services.[21] The market distinctions between RTOs and ISOs are not material here, and they can be treated as the same structure. After some experimentation, and the failure of all the alternative market designs, these organized markets under RTOs settled on the framework of an organized spot market using a bid-based, security-constrained, economic dispatch with locational marginal prices as the market foundation and substitute for the contract path model. By 2007 all the RTOs in the United States either were already using or were converting to the essential elements of this framework.

As summarized by the IEA in its review of market experience across its member countries, "Locational marginal pricing (LMP) is the electricity spot pricing model that serves as the benchmark for market design—the textbook ideal that should be the target for policy makers. A trading arrangement based on LMP takes all relevant generation and transmission costs appropriately into account and hence supports optimal investments."[22] The operation of these markets by the RTOs is the essential ingredient that allows for competition and provides open access to the transmission grid. Through these markets, the RTO provides coordination for competition, allowing participants to arrange contractual commitments without producing energy schedules and an associated power dispatch disconnected from reality. The RTO market captures the interactions among the many market participants and prices products accordingly. Furthermore, the existence of the RTO market provides the framework to create workable financial transmission rights (FTRs) that substitute for the unworkable physical transmission rights of the fictional contract path model.

20. Rajaraman and Alvarado (1998, pp. 47–54).
21. FERC (1999).
22. IEA (2007, p. 116).

The development of RTOs was impressive in its speed and scope. "These ISOs and RTOs serve two-thirds of electricity consumers in the United States and more than 50 percent of Canada's population."[23]

Outside of the organized markets served by the RTOs, which cover a large geographic area but a smaller part of the total economy and electricity system, the structure remains much as before, with vertical integration or contractual arrangements between utilities and large public providers such as the Tennessee Valley Authority in the Southeast and the Bonneville Power Administration in the Northwest.

The transformation of the organized wholesale markets was a major change in structure and market institutions that answered the unanswered questions in Order 888. The RTO experience illustrates a virtuous interplay between regulation and market design. The RTO structure is a creature of regulation, recognizing that a few critical activities essential for reliability must be centralized under monopoly control of the system operator. But the regulations and the structure are designed to internalize the problems that cannot be addressed by decentralized market decisions, dealing primarily with transmission interactions, while leaving substantial flexibility for operating and investment choices made by market participants who are responding to incentives and hold diverse views about risks and uncertainties.

Full implementation of the RTO market designs remains a work in progress. But the preservation of the traditional structure in the rest of the country presents a continuing challenge for the principles of open access. And there are more than a few difficulties in addressing the issues that arise at the seams between these various regions.

After much experimentation and lengthy debate, Congress revisited the issues of competition in the Energy Policy Act of 2005 (EPAct05).[24] This created important new authorities for FERC, such as making mandatory compliance with reliability rules that had been voluntary for decades. Voluntary compliance was plausible (but not guaranteed) under the old oligopoly, but mandatory rules were necessary in the environment of a competitive market. Furthermore, to improve planning and infrastructure development, as well as to address continuing problems of open access outside the RTOs and organized markets, FERC pursued further rulemakings to deal with transmission access and infrastructure expansion.[25]

23. See ISO/RTO Council, "ISO RTO Operating Regions" (www.isorto.org/site/c.jhKQIZPBI mE/b.2604471/k.B14E/Map.htm), especially map showing ISO-RTO operating regions.
24. P.L. 109-58.
25. FERC (2007a, 2007b, 2008a).

This substantial transformation of organized electricity markets remains controversial. Partly as a result of the reaction to the California electricity crisis of 2000–01, and partly in response to the effects of the increase in all energy prices, there has been a continuing debate about the efficacy of electricity restructuring in general and RTOs in particular. Some criticize RTOs and suggest alternative market approaches.[26] Others add to the accumulating evaluations of the benefits of RTO markets.[27] FERC will continue to be a major venue for these debates, so the issue is relevant to the agenda for action going forward.

Here there are two arguments that get entangled and confounded. First, there is the effectiveness of electricity restructuring compared to the alternative of the monopoly utility structure it was intended to replace. This is an important question on its own, and the evaluation should depend primarily on how the change affected investment decisions and innovation. Changes in operating practices with a given infrastructure are important but unlikely to dominate the comparison. The big impact should be in changed investment choices for generation, transmission, and load. The impacts here take time to unfold, and it would not be an easy matter to restore the monopoly structure.

Second is the question of how to design and regulate the market given the decision to implement electricity restructuring and change the locus of decisionmaking more toward the decentralized market. On this issue, there is now a great deal of information about what works and what does not work.

The centerpiece maintains a focus on open access and nondiscrimination in transmission services, as laid out as the objective in Order 888.[28] Outside the organized markets and RTOs, the task remains largely unfinished. Inside the organized markets and RTOs, the task is mostly complete. Moreover, every proposed alternative to the now accepted market design has been tried and found wanting. For example, a good way to put consumers in peril would be to follow recommendations such as those offered by the American Public Power Association.[29] This and similar calls to unravel RTOs would result in unworkable designs that might not even perform basic functions needed to support bilateral trading or manage congestion at least cost. The recommendation to resort to a simple contract path scheduling system, with restrictions

26. American Public Power Association (2008).

27. Harvey, McConihe, and Pope (2006). A revised 2007 version of this report is available online at www.lecg.com/files/upload/AnalysisImpactCoordinatedElectricityMkts.pdf.

28. FERC (1996a).

29. American Public Power Association (2008, pp. vii–viii).

on RTO coordinated markets, merely recycles the failed Enron initiatives of the past.[30]

The ignorance of history reflected in these reform proposals may be unwitting. But the failure to explain how bad market designs would address the problems that coordinated RTO markets handle illustrates the dangers of regulation disconnected from reality. The costs of the market design mistakes have been substantial. They could have been avoided and certainly do not have to be repeated. On the restructuring path, the emphasis should be on care in constructing regulations and market design to reflect the realities of how the system operates and how market participants respond to incentives.

The accumulated effect of these many developments reinforced the commitment to create a new regulatory structure and rules to support open access and competitive markets. In reviewing the process, FERC has emphasized both the commitment and the challenge:

> While competitive markets face challenges, we should acknowledge that competition in wholesale power markets is national policy. The Energy Policy Act of 2005 embraced wholesale competition as national policy for this country. It represented the third major federal law enacted in the last 25 years to embrace wholesale competition. To my mind, the question before the Commission is not whether competition is the correct national policy. That question has been asked and answered three times by Congress.
>
> If we accept the Commission has a duty to guard the consumer, and that competition is national policy, our duty is clear. It is to make existing wholesale markets more competitive. That is the heart of this review: to not only identify the challenges facing competitive wholesale markets but also identify and assess solutions.[31]

Going forward, the agenda for electricity markets will include at least two main strands. First, there is the unfinished task of meeting the necessary conditions for transmission open access outside the organized markets of RTOs. For instance, one litmus test is in providing open access to real-time economic dispatch for balancing markets, which is available in the organized

---

30. Hogan (1994). Available online at http://ksghome.harvard.edu/~whogan/wmkt 08041994.pdf.

31. FERC, "Chairman Kelliher's Opening Remarks at the Competition in Wholesale Power Markets Conference," February 27, 2007 (www.ferc.gov/news/statements-speeches/kelliher/2007/ 02-27-07-kelliher.asp).

RTO markets but not outside them.[32] Second, there is the continuing challenge in the organized markets of refining the market designs and addressing remaining problems that can be of particular importance in the development of electricity infrastructure.

Regulation, Markets, and Investment

Electricity restructuring has never been about deregulation or complete reliance on markets. Given current technology, it is not possible to neglect the requirements of coordination for competition. There is a need for regulation of the remaining monopoly elements of the electricity system, especially in system operations, scheduling, and dispatch. In a sense, this makes the task harder for regulators. It is one thing to oversee vertically integrated monopolies and judge the reasonableness of operations and investment decisions. It is quite another matter to fashion the rules for many different generators, transmission owners, and customers and the system operator to facilitate reasonable market decisions. And for important cases, market-driven investment may be insufficient. In these cases, some infrastructure investment decisions may still need to be made by central planners and regulators. The challenge is to formulate rules for making these decisions, rules that must work in conjunction with market choices without the unintended consequence of unraveling the market.

A fundamental task is addressing the uncertainty that confounds investment decisions. As discussed above and as emphasized in the IEA review, there are many elements of policy uncertainty that government can control, at least to an extent, and it is important to establish a workable future regime. However, certainty is not the end in itself, and a workable regime that has staying power over a long period must be compatible with the reality of how the system works and with the goal of supporting competition in wholesale power markets.

The broader government energy agenda includes a variety of issues related to climate change policy and development of new resources. The associated tax and incentive policies present their own challenges and go well beyond the scope of electricity market design and infrastructure investment rules. The framework for market design decisions falls principally within the domain of FERC and will be critical for the success of any policies. The challenges for FERC differ between those outside and inside the organized markets.

32. Chandley and Hogan (2006).

Outside the organized markets, the incentive problems are real, but they are more contained. The vertically integrated structure of the industry, full requirements service, and reliance on central planning remain, and the procedures and incentive structures of the traditional model provide the principal tools for guiding infrastructure investment. For example, there has been a continuing concern about a low level of transmission investment.[33] As directed by EPAct05, FERC issued Order 679 to promote higher rates of return on investment along with many other incentives under a regulatory model for transmission infrastructure expansion investment that will be included in the utility rate base.[34] The entire conception of the transmission expansion mechanism in Order 679 is one of traditional cost of service regulation for vertically integrated utilities. "Merchant projects are market driven while this final rule deals fundamentally with regulated transmission rates. True merchant transmission projects may play an important role in the future of transmission infrastructure development, but incentives related to, for example, return on equity and cost recovery, do not apply to merchant transmission."[35] The details are familiar from decades of utility regulation for vertically integrated utilities, but these incentive regulation details do not much confront the challenges of operating within markets.

Inside the organized markets covered by the RTOs, the required framework and tools should be different. In order to utilize the advantages of markets, it is fundamental that market participants make decisions about investments in response to incentives they perceive, bearing the risks and reaping the rewards. All these elements are necessary. Without good incentives, there is little reason to believe that market choices will be appropriate. In the face of substantial economic uncertainty, the perceptions of the market participants should govern when possible. After all, if central planners and regulators knew what to do and what investments to make, there would be no need for the market. And if the market participants do not bear the risks as well as reap the rewards, there is little hope that decisions will be good or sustainable.

Where there are market failures, some investment decisions may require central planning and regulation. For example, large-scale transmission investments may exhibit material economies of scale and scope in much the same way as assumed in the past for generation and the whole electricity system. The task for regulators, especially FERC, is to design rules for infrastructure

33. NERC (2007, p. 18).
34. FERC (2006).
35. Ibid., para. 262.

investment that address the market failures and encourage investment in a way that supports rather than undermines the market.

On the biggest issue—open access rules—the current state of market design in organized markets represents a remarkable success. The organized markets are highly regulated through the RTOs and primarily under FERC jurisdiction. However, the form of regulation through the common market design framework is materially different from that found outside the organized markets and dramatically different from the traditional model of the vertically integrated monopoly. To a reasonable approximation, control through vertical integration has been replaced by a particular form of horizontal coordination of system operations, markets, and transmission service provided as an integrated piece. Although there still are regulated utilities within organized markets that are integrated in the sense of owning their own generation, these utilities no longer control the critical central piece of transmission access. From the perspective of designing future changes in markets and incentives, therefore, vertical integration with its comforts in absorbing costs and hiding messy details is no longer available. The common market design is both more connected to the reality of how the system works and much more transparent in revealing costs and benefits.

The open access rules and common market framework are necessary for the market to support infrastructure investment, but they are not (yet) sufficient. Hence, FERC faces a continuing agenda of challenges and decisions to modify the rules and incentives to address deficiencies in the organized markets and promote investment, all while avoiding the unintended consequence of undoing the market. On this agenda, the record is troubling, but not all the record has been written. There is still the opportunity to act in time.

A broad approach to supporting reform of the electricity system would tackle the complicated jurisdictional issues between state and federal levels, and move beyond voluntary RTOs to a mandatory system. Paul Joskow has made a case for legislation to replace the Federal Power Act.[36] This would be necessary to meet all of the challenges laid out by the IEA, especially in addressing a transformation to deal with climate change and mitigation of carbon emissions. However, many of the recommendations of the IEA could be met by action at FERC under current legislative authority.

Consider two examples of infrastructure investment problems that have received a great deal of attention: resource adequacy and economic transmission

36. Paul Joskow, "Challenges for Creating a Comprehensive National Electricity Policy," speech given to the National Press Club, September 26, 2008 (www.hks.harvard.edu/hepg/Papers/Joskow _Natl_Energy_Policy.pdf).

investment. The resource adequacy questions refer primarily to the pace, type, and location of investment in assets needed to meet the baseline projected growth in demand for electricity. In principle, these assets could be anything including energy efficiency, load management, and transmission investment, but in practice the focus has been on new generation assets. The concern with economic investment in transmission is to provide adequate capacity to relieve congestion or support development of new power generation sources, primarily wind power, which must be developed far from the centers of load. In both resource adequacy and economic transmission investment cases, the problems present a challenge that appears not to be met through the current market design. The pressure has been on regulators and central planners to act now to address the long-term investment needs.

The ideal regulatory response to these problems of market failure would be to analyze the market design implementation and look first for solutions that reduce or eliminate the market failure. When this is not enough, the second response would be to create a regulatory and central planning intervention that is compatible with the market design. By contrast, the more natural response across the board of market participants, regulators, and central planners has been to make a political judgment that fixing the market design is either too hard or will not work, and to move directly to imposing a regulatory mandate that carries with it the seeds of destruction by creating new market failures and continuing pressure for further regulatory solutions.

Resource Adequacy

The problem of resource adequacy involves many dimensions and details, but the essence of the problem can be captured in a nutshell. Actual implementations of the common market framework have not quite matched the "textbook ideal" described by the IEA review. This is neither surprising nor does it require a counsel of perfection. The goal has always been to create a workable market without expecting to achieve a perfect market. However, in the case of resource adequacy and providing sufficient incentive for investment in new generation, there has been widespread concern that the incentives were not adequate.

The principal evidence cited is the so-called missing money.[37] For a variety of reasons that include price ceilings, operating procedures, and conceptual mistakes in translating theory into practice, energy prices in the electricity market have not been high enough to support investment in new

---

37. The characterization as "missing money" comes from Roy Shanker (2003).

generating plants. For example, during the nine years from 1999 through 2007, the market monitor for the PJM Interconnection (an RTO headquartered in Pennsylvania) estimates that average energy market revenue under economic dispatch for a combustion turbine peaking unit was $16,401 per MW-year compared to an average fixed cost charge of a new unit of $75,158 per MW-year. Estimates of expected net revenues going forward should be the proper benchmark, but this retrospective look at the actual revenues achieved net of variable costs is sobering and suggests a real problem in the underlying market design. The average net revenues were approximately 22 percent, 45 percent, and 63 percent of the levels needed to justify new investment in a new combustion turbine, gas-fired combined-cycle, or coal plant, respectively.[38] With these revenue gaps, it is surprising that there is as much investment in generation as has been seen.

The many analyses of the missing money problem in different parts of the country all point to essentially the same diagnosis. During times of tight supplies relative to demand—in other words, periods of relative scarcity when a normal competitive market response would be to increase price—the combined effect of the actual rules in RTO markets has been to suppress the real-time (and day-ahead) energy prices.[39] The market failure is clear. Without adequate scarcity pricing, it follows that the market incentives would not be sufficient to support investment in generation where and when it is needed.

Unfortunately, another common feature of the analyses of the missing money problem has been a widespread assumption that fixing the scarcity pricing problem would be neither politically feasible nor timely enough to be effective. (Even the terminology is politically charged, but most of the alternatives to "scarcity pricing" are equally unappealing or obscure the point.) As a result, regulators and central planners have sought other mechanisms, principally through generation capacity requirements and associated capacity markets. The experience with these capacity mandates and capacity markets has not been good. The original capacity mandates with short horizons produced disappointing results, especially in ISO New England (ISO-NE) and in PJM. This led to major efforts to reform the associated capacity markets, primarily by extending the horizon out enough years to allow new entrants to respond and get longer-term commitments from the regulators, operating through the RTO, guaranteeing certain capacity revenues separate from the energy market revenues.

---

38. PJM (2008, tables 1-3 and 3-7 through 3-9).
39. Hogan (2005).

The purpose here is not to revisit either the need for or the design of capacity markets. Absent adequate scarcity pricing, the need for something is compelling as indicated by the missing money estimates from PJM. And after much effort, the respective capacity market designs for ISO-NE and PJM are about as good as anyone has been able to construct, and current assessments promise much better performance in the future than in the past.[40]

Rather, the purpose here is to emphasize that these capacity markets are not well integrated with the rest of the market design, and they place great emphasis on the ability of central planners and regulators to make decisions in place of markets. The details of the capacity market implementations in PJM and ISO-NE are different in important ways, which itself hints that there is some disconnect between the design and the underlying fundamentals. However, these capacity market designs have in common that the RTO is assumed to be able to make good predictions about the level, type, and location of capacity needs several years into the future, and that regulators can commit to the payments for these resources while imposing the costs and the associated risks on the ultimate customers. Embedded in these analyses are assumptions about how transmission will be utilized several years into the future. The costs of the resulting capacity payments are largely socialized over time and, to an extent, over customer groups. This cost socialization makes it difficult to provide the necessary incentives for integrating demand-side efficiency investments, load management options, distributed generation, short-term operating decisions, and so on.

Faced with the substantial uncertainty about the level and location of future capacity needs, not to mention the difficulty of forecasting transmission utilization, capacity markets are markets for a regulated construct—capacity—that look much like regulated procurements in the past that led to the interest in electricity restructuring and the desire to change the locus of decisions and allocation of risks. In the face of great uncertainty and instability, regulators are now much more responsible for making investment decisions, and risks are being transferred away from investors and toward captive customers. To be sure, if regulators and central planners make good decisions, this will all work out. But if we were confident that regulators and central planners could make these decisions, we would not need electricity restructuring or the dispersed incentives in markets to prompt innovation and infrastructure investment.

40. John D. Chandley, "PJM's Reliability Pricing Mechanism: Why It's Needed and How It Works," March 2008 (www.pjm.com/documents/downloads/pjms-rpm-j-chandley.pdf). Various reports on PJM's reliability pricing model are available online at www.pjm.com.

The missing element in this story is reform of scarcity pricing to make the market implementation conform to the design framework. Significantly, there is little dispute that scarcity pricing reform would be desirable and would be compatible with the major commitment to capacity markets. Well-designed capacity markets reflect the scarcity pricing in the market, and a change in scarcity pricing rules need not change the structure of the capacity market. But better scarcity pricing would produce three principal benefits. First, better scarcity pricing would shift substantial revenues out of the capacity payments and into the energy market. This would reduce the importance of good choices by central planners and regulators, and shift the risks of being wrong away from customers and toward investors, the type of structure that is compatible with the competitive market. Second, better scarcity pricing would provide better incentives for all the innovations in the market that regulators and central planners do not anticipate and cannot plan. For example, by reducing the socialization of costs, better scarcity pricing would be much more supportive of innovation and investment on the demand side. Third, better scarcity pricing in real time would provide the last line of defense if by some misfortune the capacity investments promised by the capacity market do not arrive or are not what is needed.

Hence the choice is not between capacity markets and better scarcity pricing. For those that have implemented capacity markets, there is still a need for better scarcity pricing.[41] The argument has been that the priority was to fix the capacity market first, and better scarcity pricing could come later. The opposite priority was probably the better case, but since "later" is now, the next step is clear. In markets that have adopted capacity markets, improved scarcity pricing should be implemented to reinforce the infrastructure investment incentives and to bolster the market mechanisms for dealing with the substantial uncertainty going forward with these investments. In markets that have not implemented capacity markets, but might, better scarcity pricing should be implemented earlier rather than later.

The exact mechanism for incorporating better scarcity pricing is under discussion in various RTOs and will need support at FERC. Inevitably the pricing rules implicate other features of the electricity market, particularly the need for hedging programs for certain classes of default or core customers. In addition, better scarcity pricing is not likely to be found through trigger mechanisms that remove bid caps and rely on the exercise of market

41. Hogan (2006). Available online at http://ksghome.harvard.edu/~whogan/Hogan_PJM_Energy_Market_022306.pdf.

Figure 6-4. Operating Reserve Demand[a]

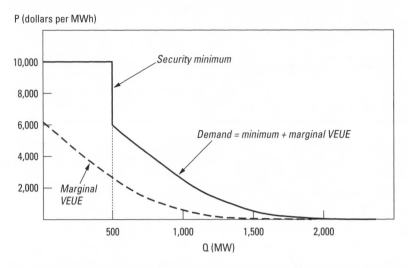

P (dollars per MWh)

Source: Hogan (2005).
a. P = price; Q = quantity.

power during periods of scarcity, which would create a headlong collision with both political sustainability and existing mandates for market power mitigation. A workable alternative would be to integrate effective operating reserve demand curves that follow on the model in New York and New England but with substantially higher prices to reflect the marginal value of expected unserved energy (VEUE) and the implied value of operating reserves, as illustrated in figure 6-4.[42]

Transmission Investment

Transmission investment rules present especially difficult problems for market design. There is a widespread concern that there has been too little investment for many years.[43] The NERC projection of transmission circuit mile growth over the next decade is approximately half of the projected growth of peak demand.[44] There has been significant transmission investment in some regions, so it is not clear if the aggregate figures reveal a pressing problem of

42. For further details, see Hogan (2005).
43. See statement of FERC chairman Joseph Kelliher (FERC 2008b, p. 13).
44. NERC (2007).

inadequate economic transmission investment. In response to the mandates of EPAct05, the U.S. Department of Energy (DOE) completed its study of transmission congestion and designated two areas, one in the east and another in the west, with significant and sustained congestion. But DOE did not address the question of whether the costs would justify investment.[45] In part, because of both uncertainty in the rules and economic uncertainty about the need for transmission, it is not an easy matter to identify economic transmission investments that are not being made.

Of course, there are many obstacles to building new transmission. Part of the motivation for the EPAct05 provisions directing DOE to identify National Interest Transmission Corridors stemmed from siting concerns and possible failures of local and state agencies in recognizing the broader regional benefits of improved transmission and wider interstate commerce. Getting siting and related environmental approvals for major transmission investments is a complicated and long process that many would like to reform. However, for our present purposes, these generic problems for transmission investment are neither new nor specific to the challenges of integrating markets and regulation.

Although market-driven merchant transmission investment is recognized as both possible in principle and occurring in practice, there is a familiar argument about a fundamental market failure in supporting transmission expansion, in particular, markets price congestion of the type identified by the DOE study. Economic investments in transmission are designed to relieve some or all of the congestion. When transmission investment is small relative to the larger market, the impact on congestion prices will be small, and the property rights created through FTRs and related benefits might be sufficient to justify the investment. However, if the investment is large enough to have a material effect on congestion prices, there could be a significant difference between the ex-ante congestion prices that point to the need for investment and the ex-post congestion prices that would define the value of the FTRs. This is a classic economy of scale argument, where monopoly provision with regulated cost recovery has a natural application. And this is the argument that markets alone may not be able to support all transmission investment.

It is easy to construct hypothetical examples where the economies of scale would apply and would be a barrier to efficient transmission investment. In this case, there is support for the regulated investment model with the asso-ciated decision rules and cost allocations. However, it would be a mistake to

45. DOE (2007).

assume that this condition applies to all or even most situations where transmission investment would be economic. Hence the challenge is to design the transmission investment rules in the organized markets to support a hybrid system with both regulated investment, where central planners make the decisions and regulators mandate cost recovery, and merchant investment, where market participants make the decisions, reap the rewards, and bear the costs.

Some of the requirements of a hybrid system seem to be clear. Property rights must be defined for the transmission investor. Cost allocation must follow the beneficiary pays principle. Decisions should defer to market choices when there is no compelling evidence of a market failure. There must be a mechanism to separate cases where regulated investment mandates would be appropriate from those where market choices should prevail.

The need for well-defined property rights is a fundamental requirement for markets and decentralized choices. If investors cannot obtain sufficient benefits from the investment, then merchant investment would be impossible to support. In the organized markets of the RTOs, the principal tool for defining and providing a workable set of property rights is award of the FTRs that would be created as a result of transmission expansions. While not perfect or complete, FTRs seem to be necessary—and in many cases, for relatively small increments, could be sufficient—to support transmission investment. A large number of small investments can have a large effect. As a result of past FERC decisions, long-term incremental FTRs are part of the design in most of the existing RTOs. This is not a particularly controversial idea.

Cost allocation according to the principle that the beneficiary of the investment pays the cost is not controversial in the abstract but is honored in the breach in practice. It is not always easy to get agreement on the identity of beneficiaries and the magnitude of the relevant benefits. Hence, while recognizing the principle, it is sometimes recommended that costs be socialized and that the allocations be balanced out over time and across portfolios of projects.[46] Innocuous as this argument may seem, it strikes at the heart of a framework that would integrate well with market decisions. This argument conflates the resulting average incidence of cost responsibility and the marginal incentive effects for investment. Under the beneficiary pays principle, the average incidence across many investments might be about the same as for a socialized cost allocation, but the incentive for each investment in trans-

46. Blue Ribbon Panel on Cost Allocation (2007, p. 7).

mission or its alternatives would be quite different under a beneficiary pays versus a socialized allocation.

Cost-benefit studies are or should be a regular part of transmission expansion evaluations.[47] A fundamental characteristic of transmission expansion is expected change in the patterns of power flow and the composition of supply, demand, and prices. For example, see the PJM analysis for the "502 Junction-Loudon Line" where the range of impacts across nineteen zones evaluated was from –$7.17 per MWh to +$8.77 per MWh, showing a material difference in the distribution of benefits.[48] If a cost-benefit study suggests that there would be net benefits, it will of necessity provide some estimate of the distribution of benefits. If that result shows the benefits are widely dispersed, it would follow from the beneficiary pays principle that cost allocation should be widely dispersed. However, this possibility should not be assumed or taken as the default assumption in favor of cost socialization. A default presumption in favor of cost socialization would undermine incentives, including the incentive to perform or evaluate the cost-benefit analysis.

Socialized cost allocation, which by definition means that regulators mandate the payment and some market participants are subsidizing others, fundamentally undermines the incentives of market investment for both transmission and its alternatives, which includes everything. It is hard to conceive of a hybrid system that widely socializes costs and still supports market-driven investments. Initially it may seem like a small and isolated problem restricted to certain complicated transmission investments, but this perception is likely to be a delusion. The forces are already in motion to extend the socialization principle to other types of infrastructure investment. For example, the DOE congestion study explicitly excluded consideration of alternatives to transmission, arguing that EPAct05 had passed that hot potato to FERC, where approval of transmission investments would have to consider alternatives to transmission investment, both in the form of generation and demand-side

47. For example, see PJM Interconnection, "Compliance Filing," docket no. ER06-1474-003, March 21, 2007 (www.pjm.com/Media/documents/ferc/2007-filings/20070321-ER06-1474.pdf) on the economic transmission planning process. See also PJM, "Business Rules for Economic Planning Process," September 2008 (www.pjm.com/~/media/planning/rtep-dev/market-efficiency/2007 0319-final-market-efficiency-business-rules.ashx).

48. PJM Planning Committee, Transmission Expansion Advisory Committee, "Market Efficiency Analysis Progress Report," April 18, 2007 (www.pjm.com/~/media/committees-groups/committees/teac/20070418-item-10-market-efficiency-analysis-prog-rep.ashx).

investments, and "FERC has committed to considering non-transmission alternatives, as appropriate, during its permit application review process."[49]

Cost socialization is not necessary for transmission investment. The beneficiary pays principle is compatible with transmission investment. For example, the Wyoming-Colorado Intertie transmission project would access distant wind resources from Wyoming and sell the energy in Colorado. The project finance model calls for the wind generators, the beneficiaries, to pay through advance commitment to purchase capacity on the line.

> As part of the Open Season process, the project sponsors had offered up to 850 megawatts of transmission capacity in a public auction. This has resulted in 585 megawatts of capacity purchase commitments from credit-worthy parties. . . . The project sponsors are optimistic that the remaining 265 megawatts of capacity will be sold. The project sponsors expect to complete the siting, permitting, and construction of the line and begin operation by mid-2013."[50]

This is a large project where beneficiaries are expected to pay.

However, FERC has shown no discernible interest in sticking to a principle when it comes to this important cost allocation principle. The FERC-approved cost allocation for ISO-NE treats 100 percent cost socialization as the default approach. Recent testimony describes one-third of costs socialized in the Southwest Power Pool, 20 percent socialization of costs for the Midwest Independent System Operator, zero percent socialization for low-voltage and 100 percent socialization for high-voltage investments in PJM.[51] The FERC-approved framework of the California ISO for "locationally constrained resource interconnection facilities," primarily for new wind generation, socializes the risk that renewable resource investments will not be adequate to pick up the cost of the transmission investment.[52] From this collection of decisions, it is not clear what the underlying recipe is, other than a recipe for trouble by driving more and more investment decisions and subsidy requests onto the regulator's agenda. Adherence to the principle of beneficiary pays seems essential for any sustainable hybrid system of transmission investment.

49. DOE (2007, p. 56994).

50. Western Area Power Administration, "Open Season a Success for Wyoming-Colorado Intertie," press release, August 26, 2008 (www.wapa.gov/newsroom/pdf/WCIOpenSeasonOutcome 82608.pdf).

51. FERC (2008b, pp. 12–13).

52. See California ISO, "Location Constrained Resource Interconnection (LCRI) Policy" (www.caiso.com/1816/1816d22953ec0.html).

The need for a hybrid system that can defer to market judgments when there is no market failure arises from the great uncertainty that surrounds all these investment choices. This uncertainty guarantees that there will be important cases where central planners and market participants have different views about the future risks and rewards. The presumed advantage of the market-based investments is that the allocation of risks is more closely linked to the decision to invest. When markets and central planners disagree, a hybrid system would require a high threshold before concluding that planners are correct and overruling market choices. And the threshold should be linked to identifying a market failure that is a barrier to entry, not just that the market failed to do what the planner preferred.

This deference to the market in a hybrid system would be closely related to the need for some mechanism for distinguishing between those infrastructure investments that would be eligible for regulated treatment and those that should be left to the market. Otherwise, everything will be a special case, and FERC will have no principled basis for deferring to the market. A particularly attractive approach follows from the early experience in Argentina and is now embodied in the transmission expansion framework proposed by the New York ISO (NYISO):

> The proposed cost allocation mechanism is based on a "beneficiaries pay" approach, consistent with the Commission's longstanding cost causation principles. . . . Beneficiaries will be those entities that economically benefit from the project, and the cost allocation among them will be based upon their relative economic benefit. . . . The proposed cost allocation mechanism will apply only if a super-majority of a project's beneficiaries agree that an economic project should proceed. The super-majority required to proceed equals 80 percent of the weighted vote of the beneficiaries associated with the project that are present at the time of the vote.[53]

This simple approach supports a hybrid scheme. Adherence to the beneficiary pays principle avoids the problems of cost socialization. The voting procedure allows for regulated transmission projects to proceed when most but not all the beneficiaries agree, while deferring to the judgment of the market when the supermajority cannot be obtained. When a small minority of beneficiaries objects, the regulatory mandate is imposed, to meet the problem of the presumed market failure. The NYISO system also embeds a principled

53. Cover letter submitted to FERC, December 7, 2007 (NYISO 2007, pp. 14–15).

answer that avoids the need to socialize the costs of most alternative investments in generation and load where the beneficiaries are usually easy to identify. If applied to transmission investment cases of a scale and scope that suggest the possibility of a market failure, there would be available a principled answer to the requests to extend the regulatory investment support to alternatives like generation and demand-side investments, which are typically small, have concentrated beneficiaries, and where there is no market failure.

The NYISO framework for regulated investments and models like the Wyoming-Colorado auction could be compatible parts of a hybrid system for transmission investment. In addition, reform of scarcity pricing would tend to increase the market price of transmission congestion, reinforcing the market incentives and beneficiary pays principle for transmission investment.

Summary

A key challenge for electricity market design and regulation is to support efficient investment in infrastructure. The IEA recommendations lead to required initiatives at FERC. Outside the organized markets, FERC faces the unfinished task of implementing and enforcing the principles of open access, which would require a change in institutions and rules similar to those of the organized markets covered by RTOs. Inside the organized markets, the continuing problem is to design rules and regulatory policies that support competitive wholesale electricity markets. A key requirement is to relate any proposed solution to the larger framework and to ask for alternatives that better support or are complementary to the market design. Many seemingly innocuous decisions appear isolated and limited but, on closer inspection, are fundamentally incompatible with and undermine the larger framework. Improved scarcity pricing and a hybrid framework for transmission investment are examples of workable solutions that seem necessary to meet the needs for a long-term approach to infrastructure investment. These are instances of the ideal regulatory response to problems of market defects and market failure. First, analyze the market design implementation and look primarily for solutions that reduce or eliminate the market failure. Second, when this is not enough, create a regulatory and central planning intervention that is compatible with the broader market design. The alternative is to frame every problem in its own terms and design ad hoc regulatory fixes that accumulate to undermine market incentives. A workable regulatory and market framework is an essential tool for anticipating unintended consequences and acting in time.

This paper draws on work for the Harvard Electricity Policy Group and the Harvard-Japan Project on Energy and the Environment. The author benefited from comments by Ralph Cavanagh, Joseph Kelliher, Elizabeth Moler, and Philip Sharp. The author is or has been a consultant on electric market reform and transmission issues for Allegheny Electric Global Market, American Electric Power, American National Power, Australian Gas Light Company, Avista Energy, Barclays, Brazil Power Exchange Administrator (ASMAE), British National Grid Company, California Independent Energy Producers Association, California Independent System Operator, Calpine Corporation, Canadian Imperial Bank of Commerce, Centerpoint Energy, Central Maine Power Company, Chubu Electric Power Company, Citigroup, Comision Reguladora de Energia (CRE, Mexico), Commonwealth Edison Company, Conectiv, Constellation Power Source, Coral Power, Credit First Suisse Boston, Detroit Edison Company, Deutsche Bank, Duquesne Light Company, Dynegy, Edison Electric Institute, Edison Mission Energy, Electricity Corporation of New Zealand, Electric Power Supply Association, El Paso Electric, GPU Inc. (and the Supporting Companies of PJM), Exelon, GPU PowerNet Pty. Ltd., GWF Energy, Independent Energy Producers Assn., ISO New England, Luz del Sur, Maine Public Advocate, Maine Public Utilities Commission, Merrill Lynch, Midwest ISO, Mirant Corporation, JP Morgan, Morgan Stanley Capital Group, National Independent Energy Producers, New England Power Company, New York Independent System Operator, New York Power Pool, New York Utilities Collaborative, Niagara Mohawk Corporation, NRG Energy, Inc., Ontario IMO, Pepco, Pinpoint Power, PJM Office of Interconnection, PPL Corporation, Public Service Electric and Gas Company, PSEG Companies, Reliant Energy, Rhode Island Public Utilities Commission, San Diego Gas and Electric Corporation, Sempra Energy, Southwest Power Pool, Texas Genco, Texas Utilities Co., Tokyo Electric Power Company, Toronto Dominion Bank, TransÉnergie, Transpower of New Zealand, Westbrook Power, Western Power Trading Forum, Williams Energy Group, and Wisconsin Electric Power Company. The views presented here are not necessarily attributable to any of those mentioned, and any remaining errors are solely the responsibility of the author.

References

American Public Power Association. 2008. "Consumers in Peril. Why RTO-Run Electricity Markets Fail to Produce Just and Reasonable Electric Rates." Washington (February).

Blue Ribbon Panel on Cost Allocation. 2007. *A National Perspective on Allocating the Costs of New Transmission Investment: Practice and Principles.* Washington, D.C.: Working Group for Investment in Reliable and Economic Electric Systems.

Chandley, John D., and William W. Hogan, 2006. "Reply Comments on Preventing Undue Discrimination and Preference in Transmission Services." Comments to

Federal Energy Regulatory Commission, docket nos. RM05-17-000 and RM05-25-000 (September 20).

Federal Energy Regulatory Commission (FERC). 1996a. "Promoting Wholesale Competition through Open Access Nondiscriminatory Transmission Services by Public Utilities; Recovery of Stranded Costs by Public Utilities and Transmitting Utilities." Order no. 888, docket nos. RM95-8-000 and RM94-7-001, final rule (April 24).

———. 1996b. "Open Access Same-Time Information System (formerly Real-Time Information Networks) and Standards of Conduct." Order no. 889, docket no. RM95-9-000, final rule (April 24, 1996).

———. 1999. "Regional Transmission Organizations." Order no. 2000, docket no. RM99-2-000 (December 20).

———. 2006. "Promoting Transmission Investment through Pricing Reform." Order no. 679, docket no. RM06-4-000, 18 CFR part 35 (July 20).

———. 2007a. "Preventing Undue Discrimination and Preference in Transmission Service." Order no. 890, docket nos. RM05-17-000 and RM05-25-000 (February 16).

———. 2007b. "Preventing Undue Discrimination and Preference in Transmission Service—Clarifying and Revising Certain Provisions of Order No. 890." Order no. 890-A, docket nos. RM05-17-001, -002 and RM05-25-001, -002 (December 28).

———. 2008a. "Preventing Undue Discrimination and Preference in Transmission Service." Order 890-B, docket nos. RM05-17-003 and RM05-25-003 (June 23).

———. 2008b. "Testimony of the Honorable Joseph T. Kelliher, Chairman, Federal Energy Regulatory Commission, before the Committee on Energy and Natural Resources, United States Senate" (July 31).

Harvey, Scott M., Bruce M. McConihe and Susan L. Pope. 2006. "Analysis of the Impact of Coordinated Electricity Markets on Consumer Electricity Charges." Paper presented at the Twentieth Annual Western Conference of the Rutgers University Center for Research in Regulated Industries, Monterey, Calif., November 20.

Hogan, William W. 1994. "An Efficient Bilateral Market Needs a Pool." Testimony at California Public Utility Commission Hearings, San Francisco (August 4).

———. 2002. "Electricity Market Restructuring: Reforms of Reforms." *Journal of Regulatory Economics* 21, no. 1: 103–32.

———. 2005. "On an 'Energy Only' Electricity Market Design for Resource Adequacy." Harvard University (September).

———. 2006. "Resource Adequacy Mandates and Scarcity Pricing ('Belts and Suspenders')." Comments submitted to the Federal Energy Regulatory Commission, docket nos. ER05-1410-000 and EL05-148-000 (February 23).

Independent System Operator New England (ISO-NE). 2008. *2007 Annual Markets Report*. Holyoke, Mass. (June).

International Energy Agency. 2007. *Tackling Investment Challenges in Power Generation in IEA Countries: Energy Market Experience*. Paris.

Myhra, David. 1984. *Whoops!/WPPSS: Washington Public Power Supply System Nuclear Plants*. Jefferson, N.C.: McFarland.

New York Independent System Operator (NYISO). 2007. "Order No. 890 Transmission Planning Compliance Filing." Docket no. OA08-13-000. Rensselaer, N.Y. (December 7).

North American Electric Reliability Corporation (NERC). 2006. "2006 Long-Term Reliability Assessment." Princeton, N.J. (October).

———. 2007. "2007 Long-Term Reliability Assessment, 2007–2016" (October).

PJM. Market Monitoring Unit. 2008. *2007 State of the Market Report*, vols. 1 and 2. Norristown, Pa.

Rajaraman, Rajesh, and Fernando L. Alvarado. 1998. "Inefficiencies of NERC's Transmission Loading Relief Procedures." *Electricity Journal* 11 (October): 47–54.

Rustebakke, Homer M., ed. 1983. *Electric Utility Systems and Practices*, 4th ed. New York: Wiley and Sons.

Shanker, Roy J. 2003. "Comments on Standard Market Design: Resource Adequacy Requirement." Comments submitted to the Federal Energy Regulatory Commission, docket RM01-12-000 (January 10).

U.S. Department of Energy (DOE). 2007. "Docket No. 2007–OE–01, Mid-Atlantic Area National Interest Electric Transmission Corridor; Docket No. 2007–OE–02, Southwest Area National Interest Electric Transmission Corridor, National Electric Transmission Congestion Report." *Federal Register* 72, no. 193 (October 5): 56992–57028.

U.S. Energy Information Administration (EIA). 2000. *The Changing Structure of the Electric Power Industry 2000: An Update*. DOE/EIA-0562(00) (October).

———. 2007. "Annual Energy Outlook (AEO) Retrospective Review: Evaluation of Projections in Past Editions (1982–2006)." DOE/EIA-0640(2006) (April).

———. 2008. *Assumptions to the Annual Energy Outlook 2008*.

U.S. Nuclear Regulatory Commission. 1979. *Annual Report—1979*. NUREG-0690.

seven

Barriers to Acting in Time on Energy and Strategies for Overcoming Them

Max H. Bazerman

The preceding chapters in this volume offer many excellent ideas on climate change; oil, transportation, and electricity policies; carbon capture and storage; and the generation of innovative energy solutions. Collectively, these papers provide the new presidential administration with a wide array of excellent policy suggestions. I will not add to this list or critique those that have been offered. Rather, I begin with the assumption that we have identified a useful, scientifically supportable agenda for changes in our energy policies. My goal is to describe the likely barriers to enacting these wise policies and present strategies for overcoming these barriers.

As noted earlier in this volume, the issue of global climate change was identified decades ago. In fact, it was first noted in the media in the 1930s, when a prolonged period of warm weather demanded explanation, yet interest in the matter disappeared as cooler temperatures returned. For the past decade, most experts have accepted climate change as a fact, making the issue difficult to ignore—yet many politicians, and the voters who elect them, have done exactly that: ignored the problem. Scientists, policymakers, and others have come up with good ideas to address climate change and the other energy issues discussed in this volume; many of the core aspects of the ideas discussed here were developed long ago. However, predictable barriers have prevented wise policies from being implemented.

The goal of this chapter is to identify and suggest ways to overcome these barriers. It explores the cognitive, organizational, and political barriers that prevent us from addressing energy problems despite clearly identified courses of action—in particular, those barriers that could hold back the policy recommendations made in the earlier chapters in this volume. It borrows from the "predictable surprises" framework that Michael Watkins and I developed to explain the human failure to act in time to prevent catastrophes.[1] It also borrows ideas from a paper of mine on cognitive barriers to addressing climate change.[2] To focus the discussion, I will treat climate change as the exemplar energy-related problem, but the ideas presented here are relevant to the enactment of wise policies across a range of issues, some of which I also discuss to demonstrate the dynamics of these barriers.

As an example of the human failure to act in time to prevent foreseeable catastrophes, Michael Watkins and I argue that our leaders had ample warning to act in time to prevent the events of September 11.[3] We note that the U.S. government knew that Islamic terrorists were willing to become martyrs for their cause and that their hatred and aggression toward the United States had increased throughout the 1990s. The American government knew that terrorists had bombed the World Trade Center in 1993, hijacked an Air France airplane in 1994 and attempted to turn it into a missile aimed at the Eiffel Tower, and attempted to simultaneously hijack eleven U.S. commercial airplanes over the Pacific Ocean in 1995. High-ranking government officials also knew that it was easy for passengers armed with small weapons to board commercial airplanes. In fact, this information was presented in many General Accounting Office reports and was identified by Vice President Al Gore's special commission on aviation security (1996).[4] Together, this information created what we called a predictable surprise and what others in this volume describe as a failure to act in time. Watkins and I argue that the failure to act in time is an unfortunately typical pattern of human behavior, one that can also be seen in the persistent failure to solve the problem of auditor independence, which contributed to the collapse of Enron, Arthur Andersen, and many other firms at the start of the millennium.[5]

Just as our government did not know how many planes the terrorists would take over on September 11 or what their targets would be, we do not

---

1. Bazerman and Watkins (2004).

2. Bazerman (2006).

3. Bazerman and Watkins (2004).

4. Office of the Vice President of the United States, *White House Commission on Aviation Safety and Security Final Report to President Clinton*, February 12, 1997 (http://www.fas.org/irp/threat/212fin~1.html).

5. Bazerman and Watkins (2004).

know which energy crisis looms largest or which will hit first. We can be confident, however, that many of the issues identified in this volume will continue to grow and that large-scale disasters will occur if we fail to act in time.

The creation and implementation of wise policy recommendations require us to anticipate resistance to change and develop strategies that can overcome these barriers. Why don't wise leaders follow through when the expected benefits of action far outweigh the expected costs from a long-term perspective? People typically respond to this question with a single explanation, a key error when explaining events.[6] This tendency to identify only one cause holds true for social problems ranging from poverty to homelessness to teenage pregnancy.[7] McGill illustrates this cognitive bias by noting that people have argued endlessly over whether teenage promiscuity or lack of birth control causes teenage pregnancy, when the obvious answer is that both cause the problem.[8] Similarly, many people seek to identify one cause of climate change when it is abundantly clear that there are multiple causes.

Enacting legislation to act in time to solve energy problems requires surmounting cognitive, organizational, and political barriers to change.[9] Efforts targeted at just one level of response will allow crucial barriers to persist. As an example, many well-intentioned organizations focus on identifying the political barriers to enacting stronger campaign finance reform in the United States. Such efforts overlook the fact that the issue of campaign finance reform is insufficiently salient in the minds of the public, and for systematic and predictable reasons. True improvements in campaign financing will require changing the way citizens think about the topic and changing the political system that continually fights against meaningful reform. And, as explored later in this chapter, current campaign financing plays a role in the political barriers to change in the realm of energy.

Drawing on this broad approach to reducing barriers to solving complex problems, the remainder of the chapter outlines three types of barriers—cognitive, organizational, and political—that confront the enactment of the wise energy recommendations in this volume. The final section moves from the identification of barriers to highlight strategies for overcoming them.

---

6. McGill (1989); Bazerman and Watkins (2004).
7. Winship and Rein (1999).
8. McGill (1989).
9. Bazerman and Hoffman (1999).

Cognitive Barriers to Acting in Time on Energy

In 2002 the Nobel Prize in Economics was awarded to a psychologist: Daniel Kahneman of Princeton University. Kahneman, together with the late Amos Tversky, created a field of study based on identifying the systematic and predictable mistakes that even very smart people make on a regular basis.[10] At the core of this field is the notion that human beings rely on simplifying strategies, or cognitive heuristics, that lead them to make predictable errors. These errors include overconfidence, anchoring of judgments, being influenced by how problems are framed, escalation of commitment, ignoring the decisions of other parties, and so forth. Scientific evidence overwhelmingly has shown that people depart from rational thought in these predictable ways, and the list of specific biases is large.[11] This literature has created scientific revolutions in economics, finance, marketing, negotiation, and medicine, among other fields, and has been popularized in many trade books.[12]

While many cognitive biases are potential barriers to the enactment of wise energy policies, three appear to be especially relevant for energy policy. First, people intuitively discount the future to a greater degree than can be rationally defended, despite their contentions that they want to leave the world in good condition for future generations. Second, positive illusions allow us to conclude that energy problems do not exist or are not severe enough to merit action—in other words, to stick our heads in the sand. Third, we interpret events in a self-serving manner, a tendency that leads us to expect others to do more than us to solve energy problems.

Discounting the Future

Would you prefer $10,000 today or $12,000 in a year? People faced with these questions often choose the former, ignoring the opportunity to earn a 20 percent return on their investment. Similarly, homeowners often fail to insulate their homes appropriately and fail to purchase energy-efficient appliances and fluorescent lighting, even when the payback would be extremely quick and the rate of return enormous. Research overwhelmingly demonstrates that people far too often use an extremely high discounting rate regarding the future, that is, they tend to focus on or overweight short-term considerations.[13]

10. Tversky and Kahneman (1974); Bazerman and Moore (2008).
11. Bazerman and Moore (2008).
12. See, for example, Thaler and Sunstein (2008); Ariely (2008).
13. Loewenstein and Thaler (1989).

Organizations also discount the future. A leading university undertook a major renovation of its infrastructure without using the most cost-efficient products from a long-term perspective.[14] Due to capital constraints on construction, the university implicitly placed a very high discount rate on construction decisions, emphasizing reduction of current costs over the long-term costs of running the building. As a result, the university passed on returns that its financial office would have been thrilled to receive on its investments. By contrast, as part of its Green Campus Initiative, Harvard University has set up a fund to finance worthwhile projects, for different colleges within the university, that may have been overlooked due to short-term budget pressures. This initiative reduces the likelihood that units of the university will make poor long-term decisions as a result of the tendency to overly discount the future.

Overdiscounting the future can create a broad array of environmental problems, from the overharvesting of the oceans and forests to the failure to invest in new technologies to respond to climate change. Herman Daly observes that many environmental decisions are made as if the earth "were a business in liquidation."[15] We discount the future the most when the future is uncertain and distant, and when intergenerational distribution is involved.[16] When people claim that we should preserve the earth, they tend to think about their descendants. But when opportunities arise that would impose environmental costs on future generations, they begin to view them as vague groups of people living in a distant time. Ackerman and Heinzerling connect the discounting of the future to driving species to extinction, the melting of polar ice caps, leaking of uranium, and failing to deal with hazardous waste. From a societal perspective, overweighting the present can be viewed not only as foolish but also immoral, as it robs future generations of opportunities and resources.[17]

Positive Illusions

The United States is likely to be substantially altered by the effects of climate change (more destructive hurricanes and the submersion of oceanfront land). Yet the George W. Bush administration repeatedly ignored opportunities to play a constructive role on climate change and failed to take steps to control or reduce the country's heavy reliance on fossil fuels. Part of the problem has been political action by the organizations most threatened by aggres-

14. Bazerman, Baron, and Shonk (2001).
15. As quoted in Gore (1993).
16. Wade-Benzoni (1999).
17. Ackerman and Heinzerling (2004); Stern (2007).

sive responses to climate change (such as auto manufacturers, oil and gas companies, and elected officials closely tied to these industries). But citizens also contribute to the problem by failing to modify their energy usage, at least until the price of gas hit $4.00 a gallon. Why have we made such egregious long-term mistakes?

One likely culprit is the existence of positive illusions about the future. In general, we tend to see ourselves, our environment, and the future more positively than is objectively the case.[18] These positive illusions have benefits, such as enhancing self-esteem, increasing commitment to action, and allowing us to persist at difficult tasks and to cope with adversity.[19] But research also shows that positive illusions reduce the quality of decisionmaking and play a role in preventing us from acting in time.[20]

While people hold a wide variety of positive illusions, two are particularly relevant to inattention to energy and climate change: unrealistic optimism and the illusion of control.[21] Unrealistic optimism is formally the tendency to believe that one's future will be better and brighter than that of other people, and better and brighter than an objective analysis would imply.[22] Undergraduates and graduate students tend to expect that they are far more likely to graduate at the top of the class, get a good job, secure a high salary, enjoy their first job, and be written up in the newspaper than reality suggests. We also assume that we are less likely than our peers to develop a drinking problem, get fired or divorced, or suffer from physical or mental problems. And we believe and act as if the changes caused by climate change will be far less significant than the scientific community predicts.

We also tend to believe that we can control uncontrollable events.[23] Experienced dice players believe that "soft" throws result in lower numbers being rolled; gamblers also believe that silence by observers is relevant to their success.[24] Such illusory behaviors result from a false belief in our control over the most uncontrollable of events. In the realm of climate change, this type of positive illusion is represented in the common expectation that scientists will invent technologies to solve the problem. Unfortunately, there is little concrete evidence that new technologies will solve the problem in time. But the

18. Taylor and Brown (1988).
19. Taylor (1989).
20. Bazerman and Watkins (2004); Bazerman and Moore (2008).
21. Bazerman, Baron, and Shonk (2001).
22. Taylor (1989).
23. Crocker (1982).
24. Langer (1975).

overestimation that new technologies will emerge serves as an ongoing excuse for the failure to act.

Egocentrism

Whose fault is climate change? As we saw in Kyoto, parties are likely to differ in their assessments of their proportionate blame and responsibility for the problem. Emerging nations blame the West for its past and present industrialization and excessive consumption. Meanwhile, the U.S. government justified its failure to ratify the agreement in part because China and India accepted little responsibility for their contribution to climate change. The United States and some other developed economies blame emerging nations for burning rainforests, overpopulation, and unchecked economic expansion.

These alternative views are consistent with the tendency to be biased in a self-serving manner, or to suffer from "egocentrism."[25] A concept related to the positive illusions described above, egocentrism refers to the tendency to make self-serving judgments regarding allocations of blame and credit, a phenomenon that in turn leads to differing assessments of what a fair solution to a problem would be.

Messick and Sentis show that we tend to first determine our preference for a certain outcome on the basis of self-interest, then justify this preference on the basis of fairness by changing the importance of the attributes affecting what is fair.[26] Thus the U.S. government might indeed want a climate change agreement that is fair to everyone, but its view of what is fair is biased by self-interest. Unfortunately, egocentrism leads all parties to believe that it is honestly fair for them to bear less responsibility for reversing climate change than an independent party would judge as fair. Thus the problem is worsened not by our desire to be unfair but by our inability to view information objectively.

Moreover, most energy issues are highly complex, lacking conclusive scientific and technological information. This uncertainty allows egocentrism to run rampant.[27] When data are clear, the mind's ability to manipulate fairness is limited; under extreme uncertainty, egocentrism is strongly exacerbated. Rawls proposed that fairness should be assessed under a "veil of ignorance"— that is, we ideally should judge a situation without knowing the role we ourselves play in it.[28] Thus, from Rawl's perspective, egocentrism describes the difference between our perceptions with and without a veil of ignorance.

25. Babcock and Loewenstein (1997); Messick and Sentis (1983).
26. Messick and Sentis (1985).
27. Wade-Benzoni, Tenbrunsel, and Bazerman (1996).
28. Rawls (1971).

Positive illusions, egocentrism, and the tendency to discount the future can have an interactive effect. After decades of insisting that the scientists are flat-out wrong, those who have strongly opposed efforts to halt climate change for self-interested reasons have changed their argument. Many no longer argue that climate change does not exist, that humans do not contribute to climate change, or that others are to blame for the problem. They now argue that it would be too costly to respond to the problem. This transition in argument—from "There is no problem" to "We are not responsible" to "It's too expensive to fix"—results in small benefits for the current generation in exchange for high costs to future generations. Regarding some details and proposals, the opponents of action on climate change may be correct, but little evidence suggests that they are interested in having their assertions tested through an objective cost-benefit analysis.

Discounting the future, positive illusions, and egocentrism are the most fundamental, innate cognitive reasons why we fail to act to address climate change. But such cognitive explanations are only part of the story.

Organizational Barriers to Acting in Time on Energy

In the United States, at least two significant organizational barriers stand in the way of the implementation of a sensible energy policy. First, the U.S. government is not currently structured in a way that would allow it to forcefully confront the nation's current energy challenges. Second, government employees often are not trained in the methods needed to implement thoughtful energy strategies.

The way in which any organization is structured affects how it collects, processes, and uses information. A common problem is that organizational "silos"—storehouses of information and resources that only certain people can access—prevent an organization from acting in time. Before September 11, no single agency or individual in the U.S. government had the specific task of managing homeland security, despite significant terrorist threats. The nation had a "terrorism czar" but no organization to support his staff's activities. Only after September 11 was the Department of Homeland Security created. Similarly, no single department is responsible for ensuring that the country is making wise decisions regarding climate change or, more broadly, energy management. While there is a Department of Energy, civilian energy is a very small part of that department's activities. No single unit is in charge of scanning the environment and collecting information on climate change, analyzing that information, and transforming it into effective policy. The United States, like other nations, developed structures for historic and institutional reasons that have not been adapted to meet current threats. Working with Congress, the new administration needs to set up structures that are better suited to the development and implementation of wise energy decisions.[29]

The second organizational barrier to acting in time has to do with what government employees in departments related to energy and the environment have been trained and rewarded to do. Both the U.S. Environmental Protection Agency (EPA) and the Department of the Interior have developed regulatory regimes over the past thirty-five years that are based on a command-and-control structure. Government employees have been trained to penalize corporations and landowners that act against established standards.[30] Some presidential administrations have set tougher standards and enforced them more strictly than others, but the nature of the regulatory structure has remained. Once regulations are created, government employees adopt a compliance mindset that attenuates the creative search for more economically and environmentally efficient choices that might deviate from the standard.[31]

In response to such inefficiency, two fascinating programs were developed in the 1990s specifically to help create wiser tradeoffs between environmental and private interests regarding environmental issues: Project XL and Habitat Conservation Plans (HCPs). As part of the Clinton administration's goal of "reinventing government," both programs offered relief on specific regulations in return for overall superior environmental performance.

In May 1995 the EPA introduced Project XL (eXcellence and Leadership) to foster cooperation with regulated companies in the development of more cost-efficient and effective environmental protection. Project XL gave corporations greater flexibility in achieving the government's environmental objectives, provided they met current regulatory standards. Specifically, companies received regulatory relief in return for "superior environmental performance" as compared to a baseline. Overall, proposals under Project XL had to produce private and regulatory cost savings, be supported by stakeholders, and avoid shifting safety risks to other potentially affected parties.

Unfortunately, Project XL had only limited success. The number of projects approved and implemented fell short of the EPA's initial objectives. Nearly 27,000 facilities released hazardous and toxic materials in 2000, yet only three XL projects were proposed that year, the final year of the Clinton administra-

29. Ogden, Podesta, and Deutch (2008).
30. Hoffman, Bazerman, and Yaffee (1997).
31. Tenbrunsel and others (1997).

tion.[32] As part of its tendency to unilaterally weaken environmental regulation and enforcement rather than encouraging cooperation between environmental and private interests, the Bush administration closed down Project XL to new proposals in 2003. In the end, the number of terminated or inactive XL projects was greater than the number of projects implemented and completed.

Congress introduced HCPs in 1982 as an antidote to the shortcomings of the Endangered Species Act (ESA). Enacted in 1972, the ESA prohibits the "take" of any federally listed animal or plant species considered "endangered" or "threatened" from public or private land. To "take," as defined in the ESA, means to "harass, harm, pursue, hunt, shoot, wound, kill, trap, capture, collect or attempt to engage in such conduct."[33] The ESA's prohibition on taking protected species can impose severe land use restrictions that, for many landowners and developers, appear to violate sacrosanct private property rights without just compensation. The ESA set up an adversarial conflict between the government and landowners that created incentives contrary to the objectives of species protection. For instance, some landowners have destroyed species habitat, choosing to "shoot, shovel, and shut up," in the words of one landowner, for fear of government intervention.[34]

HCPs were created to allow landowners to negotiate compliance with the ESA while retaining control of their land. HCPs allow the "incidental taking" of endangered species in exchange for a commitment by the landowner to provide enhanced protection for the species over a longer time horizon. For the first ten years of the program, HCPs saw little use.[35] In 1995 Secretary of the Interior Bruce Babbitt began to promote HCPs as a useful tool, and large-scale plans (in excess of 1,000 acres) were proposed.

Just as Project XL failed to catch on, HCPs have not been fully accepted as a new form of cooperation between public and private interests. While the HCP process still exists, its use has been disappointing. The departure of Clinton and Babbitt reduced the creativity that had existed within the Department of the Interior. In addition, both HCPs and Project XL ended up being far more complex and bureaucratic than their creators envisioned.

At their core, Project XL and HCPs were wise environmental policies. They were capable of promoting the kinds of creative tradeoffs that we teach graduate and executive students to explore. Why were these excellent ideas so difficult

---

32. Hoffman and others (2002).

33. See U.S. Fish and Wildlife Service, Endangered Species Program, "Endangered Species Act of 1973, as Amended through the 108th Congress. Sec. 3" (www.fws.gov/endangered/esa/sec3.html).

34. Crismon (1998).

35. Noss, O'Connell, and Murphy (1997).

to implement? Only by identifying the core, taken-for-granted beliefs of regulatory institutions such as the EPA can we understand the persistence of inefficient regulatory design and the barriers to acting in time on energy. To allow the next administration to implement the kind of energy and environmental policies needed to act in time, we must change how individuals think and how institutions guide that thinking.

Institutions are composed of the laws, rules, protocols, standard operating procedures, and accepted norms that guide organizational action.[36] Members of institutions who adopt these laws, protocols, and norms gradually come to behave by force of habit. In turn, habit creates resistance to change and leads institutional members to reject new forms of regulatory policy. Project XL and HCPs represented revolutionary change that was met with resistance caused by years of institutional inertia.

Moreover, for cooperative regulatory reform to work, trust between parties is essential.[37] Voluntary information sharing and regulatory flexibility are at the heart of Project XL and HCPs, yet both are anathema to many bureaucratic departments of government. Indeed, the United States has traditionally shunned creating the kind of cooperative regulations that are common in Asian and European countries. As one editorialist quipped about the EPA, "Does anyone truly believe that any government bureaucracy—especially one so deeply suspicious of the regulated community, an agency that measures its worth by its annual tally of convictions of environmental miscreants—would actually be willing to bargain away its birthright?"[38] Giving up control, as well as the idea of "negotiating" environmental improvements, may appear to some regulators to be contrary to their mandate of protecting the environment.

Furthermore, regulators may resist the shift from command-and-control to cooperative regulation for fear of losing responsibilities, power, and competence. Organizational confusion and turf wars between rival departments can be the inevitable result. Anne Kelley, a staff member of the EPA's New England Region, had this to say about reinvention efforts and Project XL: "I represented a tiny office that came [to the EPA] begging for open-mindedness, but unfortunately most in the agency locked arms against reinvention."[39] In several Project XL negotiations, companies complained that EPA staff assigned to the project lacked the authority needed to make decisions and the resources needed to support the project adequately. In addition, gov-

36. Scott (1995).
37. Ruckelshaus (1996).
38. Harris (1996, p. 4).
39. Hoffman and others (2002, p. 838).

ernment scientists were not given the negotiation training they needed to successfully hammer out complex deals with business interests.[40]

In sum, beyond identifying wise policies, the new administration must anticipate and address those aspects of government organizations that will prevent the successful implementation of new ideas aimed at acting in time to solve energy problems.

Political Barriers to Acting in Time on Energy

The failure of the U.S. government to pass meaningful and sufficient campaign finance reform laws perpetuates a system in which money corrupts the potential for an intelligent decisionmaking process on energy policy. Well-funded and well-organized special interest groups—concentrated constituencies intensely concerned about a particular issue—have disproportionate influence on specific policies at the expense of millions who lack a strong voice on that issue.

Experienced at subverting good ideas, leading organizations from the automotive, coal, and oil industries (for example, ExxonMobil) have succeeded in distorting energy politics and keeping the United States from implementing wise practices regarding climate change. These special interest groups lobby elected officials against acting in time to prevent climate change and will continue to try to do so. They stall reforms by calling for more thought and study or by simply donating enough money to the right politicians so that wise legislation never even comes to a vote. Their efforts effectively turn Congress and the president away from the challenge of making wise energy decisions.

Special interest groups that want to block better policies use a key tool: obfuscation. The tobacco industry successfully relied on obfuscation to block regulation for decades. The industry knew about the harms of cigarette smoking, and then second-hand smoke, long before the public did. To avoid or slow down antismoking measures, the tobacco industry created confusion about the effects of smoking through misleading advertising and industry-funded "research." Similarly, vocal representatives of the coal, oil, and automobile industries first denied the existence of climate change, and then obfuscated the role that people play in the problem and now the magnitude of the problem. Obfuscation works: as explained earlier, people are less willing to invest in solving problems perceived to be uncertain.

40. Ibid.

In part due to industry-sponsored obfuscation, any elected official who supports measures aimed at combating climate change can expect constituents to question the wisdom of incurring the substantial costs of action, especially if those costs include new taxes on SUVs, gasoline, and so on. Public officials are faced with the dilemma of imposing costs (such as gas taxes) on the current generation for a problem that is out of focus for many constituents. Without knowledge of the potentially disastrous long-term effects and costs of climate change, the public is unlikely to enthusiastically endorse these short-term costs. The natural human impulse, as I described earlier, is to discount the future. This uninformed preference keeps the public from endorsing the actions of politicians who accept the need to inflict small costs in the present to avoid a future catastrophe. As recently as the 2008 primary season, politicians clamored to offer the public lower gas prices—proposals rejected by economists and scientists alike as fundamentally unsound. As with other issues, U.S. energy policy will be compromised if we do not also address special interest group politics and enact meaningful campaign finance reform.

Overcoming Barriers

While I have separately addressed cognitive, organizational, and political barriers to implementing wise energy policy, it is important to recognize that the processes that prevent wise policy formulation are interconnected. Any plan to act in time on energy must be cognizant of the forces that will work against change. Responding to cognitive barriers while ignoring organizational and political barriers will not solve the problem. Similarly, political or organizational change will not occur as long as leaders and citizens are affected by the biases documented here.

This final section identifies a series of principles aimed at attempting to overcome barriers to the implementation of wise energy policies. Rather than trying to develop new policies, it suggests ways to improve the odds that policies that have already been identified will succeed.

Principle 1: The 2009 U.S. Administration Should Identify and Educate the Public about Energy Policies That Make Wise Tradeoffs across Issues.

The continuation of many policies can often obscure the fact that they no longer—if they ever did—entail wise tradeoffs. A perfect example concerns current practice in the United States with regard to organ donation. Over 30 percent of the 40,000 Americans waiting for an organ transplant are likely to die before an organ is found—yet many Americans are buried each year with potentially lifesaving organs intact. The majority of European nations have significantly increased organ donation rates by encouraging citizens to accept a simple switch in mindset.[41] Rather than handing out donor cards to those who consent to donate, the government gives citizens the right to opt out of donating their organs; in other words, citizens who do not object are automatically assumed to be organ donors. This change in the default policy has had an enormous effect, more than doubling the effective donation rate. Influenced by those who argue, in essence, that the sanctity of the human body is more sacred than the lives of those awaiting organs, U.S. politicians have not instituted this change.

In my opinion, favoring donors over recipients is a poor tradeoff that costs too many lives—and perhaps, one day, yours or mine. I think most citizens would agree with this perspective if this tradeoff were made clear to them. Few policies are as inefficient as the U.S. organ donation system, but others are quite troubling.[42] Why do we tolerate a legal system that discourages pharmaceutical companies from developing drugs and vaccines that would ease suffering and save lives? Why do we subsidize tobacco farmers to grow a crop directly responsible for close to 430,000 U.S. deaths each year? Why have we depleted many of the world's most valuable and abundant fishing basins?

And in the realm of energy policy, why haven't we implemented tougher mileage requirements for cars? Why haven't we invested more in educating the public about ways to reduce energy usage that would save them money? Why haven't we developed programs that allow companies to make wise long-term decisions as they build new plants, decisions that would be good for the firm, good for energy conservation, and good for the environment?

My colleagues and I have argued that many of these failures occur because losses loom larger than gains in the minds of citizens and politicians.[43] Consider that most policy changes create both gains and losses. Society often misses opportunities for wise tradeoffs—those in which gains significantly exceed losses for all parties—by failing to implement policies that have some

41. Johnson and Goldstein (2003).
42. Bazerman, Baron, and Shonk (2001).
43. Ibid. See also Kahneman and Tversky (1979).

costs, even when the gains are far greater. When losses loom larger than gains, we fail to act in time to make decisions that would create a net benefit.

Principle 2: The 2009 U.S. Administration Should Seek Near-Pareto Improvements and Communicate That Decisions Will Be Made to Maximize Benefits to Society Rather Than to Special Interest Groups.

Rather than fighting over the importance of addressing energy issues, elected officials should work together to identify wise tradeoffs on energy issues. A perfect wise trade would create a policy change that economists call a "pareto improvement"—a change that would make some people better off and no one worse off. Unfortunately, in a country of 300 million citizens, true pareto improvements are rare to nonexistent in government policymaking. Most changes will require some sacrifices from some members of society. Thus, in most policy domains, we should seek what economist Joseph Stiglitz calls "near-pareto improvements": changes that create vast benefits for some and comparatively trivial losses for others, or hurt only a small, narrowly defined special interest group—often, a group that has already manipulated the political process to its advantage.[44]

I hope that the new administration will agree with Stiglitz's argument that "if everyone except a narrowly defined special interest group could be shown to benefit, surely the change should be made."[45] The new administration should make this principle transparent. Obviously, advocating policies connected to campaign finance reform (including public financing of campaigns) would be consistent with this principle and, in turn, would promote the enactment of a wiser energy policy.

Principle 3: The 2009 U.S. Administration Should Seek Energy Policies That Make Sense Even If Climate Change Is Less of a Problem Than Best Current Estimates Suggest.

Some changes to energy policy will be difficult to pass into law and to implement due to uncertainty about the future. Others should be easy. Some actions we could take to reduce greenhouse gases could be beneficial for reasons other than reducing climate change; for example, improvements in energy efficiency could be cost-effective in their own right. Most of us make poor tradeoffs across time due to our tendency to discount the future. We should seek to identify the multitude of policies that would create "no regrets," regardless of

44. Stiglitz (1998).
45. Ibid.

how uncertainties play out. As many politicians have noted, a no-regrets strategy would be beneficial even if climate change turns out to be a lot of hot air.

Principle 4: The Next U.S. Presidential Administration Should Identify a Series of Small Changes (Nudges) That Significantly Influence the Behaviors of Individuals and Organizations in a Positive Direction without Infringing on Personal Liberties.

In their book *Nudge*, Richard Thaler and Cass Sunstein advocate a strategy that they call "libertarian paternalism."[46] Essentially, using the cognitive biases described earlier in this chapter, they develop policy suggestions that account for how people actually make decisions and "nudge" people toward making wiser decisions. Thaler and Sunstein's strategies are "paternalistic" because they manipulate people to act according to the preferences of the policy designers; the strategies are "libertarian" because the policies do not limit individual freedom.

Citing an example of a beneficial nudge, Thaler and Sunstein note that the U.S. Congress responded to the Chernobyl disaster by passing the Emergency Planning and Community Right to Know Act. This act required companies that pollute to create a Toxic Release Inventory. While the act was primarily a bookkeeping measure that had little regulatory teeth, it succeeded in achieving large reductions in toxic releases. How did it succeed? According to Thaler and Sunstein, environmental groups used the Toxic Release Inventory to produce and publicize "blacklists" of polluters. Fearful of negative publicity and falling stock prices, polluters cleaned up their act to avoid being blacklisted, resulting in a phenomenon that Thaler and Sunstein call a "social nudge."

In an example of a small positive nudge in the realm of energy policy, Thaler and Sunstein criticize the format of the current fuel economy sticker that is required on all new automobiles.[47] They argue that the sticker would have a much greater impact in reducing gas consumption by providing information more relevant to consumer use. The currently required sticker includes expected miles per gallon on the highway and in the city, plus some technical information that is likely ignored by 95 percent of customers as well as comparison information to similar cars in small print. Thaler and Sunstein recommend a sticker that provides highway and city estimates, cuts the technical information, puts the comparison data in an easy-to-read chart, and provides an estimate of the annual fuel cost of driving the car. My intuition is that Thaler and Sunstein's sticker would have a far greater impact on gasoline consumption

46. Thaler and Sunstein (2008).
47. Ibid.

than the current sticker. And, other than the manufacturers of fuel-inefficient vehicles, it is hard to imagine who would be against the new sticker.

Another example of a beneficial nudge, this one described in the *New York Times*, comes from the town of Hove, England, which placed "smart meters" to chart electricity usage in citizens' homes.[48] These meters provide residents with information about how much electricity their homes are currently using: turn on a high-use appliance and the meter jumps. This simple feedback has not only made Hove residents aware of their energy use, but it has also motivated them to conserve energy. The program has been a big success, and the British government is considering placing some version of the smart meter in all 46 million homes in the United Kingdom.

Principle 5: When Discounting of the Future Creates an Insurmountable Barrier to the Implementation of Wise Policies, Consider Implementation on a Mild Delay.

Many wise energy policies require people to make a small-to-medium current sacrifice in return for larger benefits in the future (or to avoid larger future harms). Todd Rogers and I have shown that such proposals tend to fail because people overweight the immediate costs of implementation.[49] Laibson's work on hyperbolic discounting shows that people's discount function is not linear but rather in a shape that resembles a hyperbola.[50] In other words, while most people would prefer $5 today over $6 tomorrow, most would also prefer $6 in 31 days over $5 in 30 days.[51] Applied to the policy arena, Rogers and I show that people are more likely to support energy policies that have initial costs and long-term benefits when the policies will be implemented in the future—even in the near future—rather than today. While a small delay may create inefficiency, we find large increases in support even for slightly delayed policies. Effectively, the small delay gets people to look beyond the steep slope of the loss function created by the current costs.

Conclusion

As I write this chapter, I am sure that others can develop far better ideas for overcoming cognitive, organizational, and political barriers to implementing

---

48. Elisabeth Rosenthal, "Trying to Build a Greener Britain, Home by Home," *New York Times*, July 20, 2008, p. 6.

49. Rogers and Bazerman (2008).

50. Laibson (1994).

51. Frederick, Loewenstein, and O'Donoghue (2002).

wise energy policies. My goal has been to focus our attention on the need to confront these barriers to change and to suggest concrete steps to overcome them. If this chapter prompts others to develop additional and better strategies to overcome barriers to wise energy policies, it will have served its purpose.

References

Ackerman, F., and L. Heinzerling. 2004. *Priceless.* New York: New Press.

Ariely, D. 2008. *Predictably Irrational: The Hidden Forces that Shape Our Decisions.* New York: HarperCollins.

Babcock, L., and G. Loewenstein. 1997. "Explaining Bargaining Impasse: The Role of Self-Serving Biases." *Journal of Economic Perspectives* 11, no. 1: 109–26.

Bazerman, Max H. 2006. "Climate Change as a Predictable Surprise." *Climatic Change* 77 (July): 179–93.

Bazerman, Max H., J. Baron, and K. Shonk. 2001. *You Can't Enlarge the Pie: Six Barriers to Effective Government.* New York: Basic Books.

Bazerman, Max H., and A. J. Hoffman. 1999. "Sources of Environmentally Destructive Behavior: Individual, Organizational, and Institutional Perspectives." In *Research in Organizational Behavior,* vol. 21, edited by R. I. Sutton and Barry M. Staw, pp. 39–79. Stamford, Conn.: JAI Press.

Bazerman, Max H., and Don A. Moore. 2008. *Judgment in Managerial Decision Making.* 7th ed. New York: John Wiley and Sons.

Bazerman, Max H., and M. D. Watkins. 2004. *Predictable Surprises: The Disasters You Should Have Seen Coming and How to Prevent Them.* Harvard Business School Press.

Crismon, S. 1998. "North Carolina Red-Cockaded Woodpecker Habitat Conservation Plan and Safe Harbor." In *Improving Integrated Natural Resource Planning: Habitat Conservation Plans.* Knoxville, Tenn.: National Center for Environmental Decision-Making Research.

Crocker, J. 1982. "Biased Questions in Judgment of Covariation Studies." *Personality and Social Psychology Bulletin* 8, no. 2: 214–20.

Frederick, S., G. Loewenstein, and T. O'Donoghue. 2002. "Time Discounting and Time Preference: A Critical Review." *Journal of Economic Literature* 40, no. 2: 351–401.

Gore, Al. 1993. *Earth in the Balance.* New York: Penguin Books.

Harris, P. 1996. "They Just Don't Get It." *Environmental Management Today* 7: 4.

Hoffman, Andrew J., Max H. Bazerman, and Steven L. Yaffee. 1997. "The Endangered Species Act and the U.S. Economy." *Sloan Management Review* 39 (Fall): 59–73.

Hoffman, A. J., and others. 2002. "Cognitive and Institutional Barriers to New Forms of Cooperation on Environmental Protection." *American Behavioral Scientist* 45, no. 5: 820–45.

Johnson, Eric J., and Daniel Goldstein. 2003. "Do Defaults Save Lives?" *Science* 302, no. 5649: 1338–39.

Kahneman, Daniel, and Amos Tversky. 1979. "Prospect Theory: An Analysis of Decision under Risk." *Econometrica* 47: 263–91.

Laibson, D. I. 1994. "Essays in Hyperbolic Discounting." PhD dissertation, Massachusetts Institute of Technology.

Langer, E. J. 1975. "The Illusion of Control." *Journal of Personality and Social Psychology* 32, no. 2: 311–28.

Loewenstein, George, and Richard H. Thaler. 1989. "Anomalies: Intertemporal Choice." *Journal of Economic Perspectives* 3: 181–93.

McGill, A. L. 1989. "Context Effects in Judgments of Causation." *Journal of Personality and Social Psychology* 57, no. 2: 189–200.

Messick, D. M., and K. P. Sentis. 1983. "Fairness, Preference, and Fairness Biases." In *Equity Theory: Psychological and Sociological Perspectives*, edited by D. M. Messick and K. S. Cook, pp. 61–94. New York: Praeger.

———. 1985. "Estimating Social and Nonsocial Utility Functions from Ordinal Data." *European Journal of Social Psychology* 15, no. 4: 389–99.

Noss, Reed F., Michael A. O'Connell, and Dennis D. Murphy. 1997. *The Science of Conservation Planning: Habitat Conservation under the Endangered Species Act.* Washington: Island Press.

Ogden, Peter, John Podesta, and John Deutch. 2008. "New Strategy to Spur Energy Innovation." *Issues in Science and Technology* 24 (Winter): 35–44.

Rawls, J. 1971. *A Theory of Justice.* Harvard University Press.

Rogers, Todd, and Max H. Bazerman. 2008. "Future Lock-In: Future Implementation Increases Selection of 'Should' Choices." *Organizational Behavioral and Human Decision Processes* 106, no. 1: 1–20.

Ruckelshaus, William D. 1996. "Restoring Public Trust in Government: A Prescription for Restoration," Webb Lecture Series. Washington: National Academy of Public Administration.

Scott, W. R. 1995. *Institutions and Organizations.* London: Sage Publications.

Stern, Nicholas. 2007. *The Economics of Climate Change: The Stern Review.* Cambridge University Press.

Stiglitz, Joseph. 1998. "The Private Uses of Public Interests: Incentives and Institutions." *Journal of Economic Perspectives* 12, no. 2: 3–22.

Taylor, S. E. 1989. *Positive Illusions: Creative Self-Deception and the Healthy Mind.* New York: Basic Books.

Taylor, S. E., and J. D. Brown. 1988. "Illusion and Well-Being: A Social Psychological Perspective on Mental Health." *Psychological Bulletin* 103, no. 2: 193–10.

Tenbrunsel, Ann E., and others. 1997. "The Dysfunctional Aspects of Environmental Standards." In *Environment, Ethics, and Behavior: The Psychology of Environmental Valuation and Degradation,* edited by M. H. Bazerman and others, pp. 105–21. San Francisco: New Lexington Press.

Thaler, Richard H., and Cass R. Sunstein. 2008. *Nudge: Improving Decisions about Health, Wealth, and Happiness.* Yale University Press.

Tversky, Amos, and Daniel Kahneman. 1974. "Judgment under Uncertainty: Heuristics and Biases." *Science* 185, no. 4157: 1124–31.

Wade-Benzoni, Kimberly A. 1999. "Thinking about the Future: An Intergenerational Perspective on the Conflict and Compatibility between Economic and Environmental Interests." *American Behavioral Scientist* 42, no. 8: 1393–405.

Wade-Benzoni, Kimberly A., Ann E. Tenbrunsel, and Max H. Bazerman. 1996. "Egocentric Interpretations of Fairness in Asymmetric, Environmental Social Dilemmas: Explaining Harvesting Behavior and the Role of Communication." *Organizational Behavior and Human Decision Processes* 67, no. 2: 111–26.

Winship, C., and M. Rein. 1999. "The Dangers of 'Strong' Causal Reasoning in Social Policy." *Society* 36, no. 5: 38–46.

Contributors

Laura Diaz Anadon manages the Energy Research, Development, Demonstration, and Deployment Policy project at the Harvard Kennedy School. Anadon holds a Ph.D. in chemical engineering from the University of Cambridge and has done research at DuPont, Bayer Pharmaceuticals, and Johnson Matthey; she has also worked as a financial consultant.

Max H. Bazerman is the Straus Professor at the Harvard Business School and is formally affiliated with the Harvard Kennedy School, the Psychology Department, and the Harvard Program on Negotiation. He is the author, coauthor, or coeditor of sixteen books (including *Negotiation Genius*, with Deepak Malhotra) and over two hundred research articles and chapters. In 2006 Bazerman received an honorary doctorate from the University of London (London Business School), the Kulp-Wright Book Award from the American Risk and Insurance Association for *Predictable Surprises* (with Michael Watkins), and the Life Achievement Award from the Aspen Institute's Business and Society Program. In 2008 he received the Distinguished Educator Award from the Academy of Management.

Kelly Sims Gallagher is director of the Energy Technology Innovation Policy research group and adjunct lecturer at the Harvard Kennedy School. She holds an M.A.L.D. and Ph.D. in international relations from the Fletcher School at Tufts University. Current research interests are in U.S. and Chinese energy and climate policy, and policy for energy technology innovation, including technology transfer issues. Her book, *China Shifts Gears:*

Automakers, Oil, Pollution, and Development, is available from MIT Press and Tsinghua University Press.

William W. Hogan is Raymond Plank Professor of Global Energy Policy at the Harvard Kennedy School. He is research director of the Harvard Electricity Policy Group, which is examining alternative strategies for a more competitive electricity market. Hogan was a member of the faculty of Stanford University, where he founded the Energy Modeling Forum, and he is a past president of the International Association for Energy Economics. Current research focuses on major energy industry restructuring, network pricing and access issues, and market design. Hogan received his undergraduate degree from the U.S. Air Force Academy and his Ph.D. from the University of California at Los Angeles. Selected papers are available on his website, www. whogan.com.

John P. Holdren was nominated as Assistant to President Barack Obama for Science and Technology in December 2008. He was Teresa and John Heinz Professor of Environmental Policy and director of the Program on Science, Technology, and Public Policy at the Harvard Kennedy School, and also professor of environmental science and policy in Harvard's Department of Earth and Planetary Sciences prior to this appointment. He was president and director of the Woods Hole Research Center and cochair of the independent, bipartisan National Commission on Energy Policy. Trained in aerospace engineering and theoretical plasma physics at MIT and Stanford, Holdren cofounded in 1973 the program in energy and resources at the University of California, Berkeley. He is a member of the National Academy of Sciences, National Academy of Engineering, American Academy of Arts and Sciences, Council on Foreign Relations, and served as president of the American Association for the Advancement of Science.

Henry Lee is the Jassim M. Jaidah Family Director of the Environment and Natural Resources Program and lecturer in public policy at the Harvard Kennedy School. During the past twenty years, his research has focused on policies to address energy security and global environmental threats.

Daniel P. Schrag is professor of earth and planetary sciences and environmental engineering at Harvard University and the director of the Harvard University Center for the Environment. Schrag studies climate and climate change over the broadest range of earth history. He has examined changes in ocean circulation over the last several decades, with particular attention to El Niño and the tropical Pacific. He has worked on theories for Pleistocene ice age cycles and has contributed to the development of the Snowball Earth hypothesis. Schrag is currently working on technological approaches to mitigating the effects of human-induced climate change. Among various honors, Schrag was awarded a MacArthur Fellowship in 2000. He came to Harvard in 1997 after teaching at Princeton and studying at the University of California, Berkeley, and at Yale.

Index